T0184876

Innovations in Science Education and Technology

Volume 22

Series editor
Cohen, Karen C.
Weston, MA, USA

As technology rapidly matures and impacts on our ability to understand science as well as on the process of science education, this series focuses on in-depth treatment of topics related to our common goal: global improvement in science education. Each research-based book is written by and for researchers, faculty, teachers, students, and educational technologists. Diverse in content and scope, they reflect the increasingly interdisciplinary and multidisciplinary approaches required to effect change and improvement in teaching, policy, and practice and provide an understanding of the use and role of the technologies in bringing benefit globally to all.

More information about this series at http://www.springer.com/series/6150

Anat Yarden • Stephen P. Norris
Linda M. Phillips

Adapted Primary Literature

The Use of Authentic Scientific Texts
in Secondary Schools

 Springer

Anat Yarden
Weizmann Institute of Science
Rehovot, Israel

Stephen P. Norris (deceased)
Educational Policy Studies
University of Alberta
Edmonton, AB, Canada

Linda M. Phillips
Canadian Centre for Research
    on Literacy
University of Alberta
Edmonton, AB, Canada

ISSN 1873-1058   ISSN 2213-2236   (electronic)
Innovations in Science Education and Technology
ISBN 978-94-017-7826-8   ISBN 978-94-017-9759-7   (eBook)
DOI 10.1007/978-94-017-9759-7

Springer Dordrecht Heidelberg New York London
© Springer Science+Business Media Dordrecht 2015
Softcover reprint of the hardcover 1st edition 2015

Printed on acid-free paper

Springer Science+Business Media B.V. Dordrecht is part of Springer Science+Business Media (www.springer.com)

This book is dedicated to Stephen Norris, his memory, his legacy, and his family and friends.

When we three (Anat, Steve and Linda) agreed to work together in 2010 it was an exciting time for us. We bubbled with ideas and looked forward very much to taking on this book. We planned and met at meetings, talked over breakfasts and dinners and skyped numerous times to clarify our thoughts and polish our wording.

On February 18, 2014, all of that changed with Steve's sudden and untimely passing. Steve is well known internationally for his leading work on reading scientific texts, scientific literacy, scientific knowledge and reasoning, the public understanding of science as well as his work on test validation and critical thinking. He has an outstanding research and publication record with numerous high quality publications in the form of books, peer-reviewed articles, and chapters. His work has been funded by the Social Sciences and Humanities Research Council of Canada for the duration of his academic career. He was a Tier 1 Canada

Research Chair in Scientific Literacy and the Public Understanding of Science, notably he was the sole scholar in Canada awarded a research chair in the field of science education. Within Canada he was a national leader for the five centers for research in youth, science teaching and learning funded by the Natural Sciences and Engineering Research Council of Canada. Steve's work is used internationally and he has given willingly of his time and expertise.

Jonathan Osbourne, a colleague and friend, said it well: At his heart, Steve Norris was what all communities need—a critical friend. Watching a presentation by him was to observe a model of clarity both in the deliberate thoughtful manner it was presented and in the depth of thought that had gone into his arguments and questions. He was somebody who recognized that the first duty of an intelligent man is to state the obvious and to ask the hard questions that others had avoided. In doing so, he enriched our community and advanced our thinking. He was also an individual of great wit and charm who took profound interest in helping and supporting all. He passed away doing what he loved, outdoors snowshoeing with his wife and friends. For him it was swift. For those of us who knew him it is another rent in the fabric of life. For our science community it is a great loss.

Steve was engaged completely with the writing of this book with his two friends and colleagues at the time of his death. He had the path clearly specified and we took up the challenge and continued to complete this book without his thoughtful and helpful comments. We trust that we have served him well. He saw innovative possibilities for the use and adoption of adapted primary literature (APL) and was striving to advance the idea in various academic platforms, including the writing of this book. We hope this book will enable others to realize Steve's vision and grasp the APL idea for promoting science education through the reading of authentic scientific texts.

We are indeed blessed to have known Steve and doubly blessed to have worked and walked many paths with him.

Anat Yarden and Linda Phillips

Stephen P. Norris
1949–2014

Contents

About the Authors

Dr. Stephen P. Norris* was Professor and Canada Research Chair in Scientific Literacy and the Public Understanding of Science at the University of Alberta, Canada. He has undergraduate degrees in Physics and in Science Education, a master's degree in Science Education, and a Ph.D. in the Philosophy of Education. He has published extensively in philosophy of education and science education for over three decades. Steve has a longstanding and international reputation for his work on reading scientific texts, scientific literacy, scientific knowledge and reasoning, as well as the public understanding of science as well as his work on testing and critical thinking. During the last 20 years he has published extensively on interpreting scientific text. His most recent books are: a co-authored volume with Linda Phillips and John Macnab, *Visualization in Mathematics, Reading and Science Education* published by Springer; and an edited volume, *Reading for Evidence and Interpreting Visualizations in Mathematics and Science Education*, published by Sense.

Dr. Linda M. Phillips is Centennial Professor and Director of the Canadian Centre for Research on Literacy at the University of Alberta. She holds two undergraduate degrees and a master's from Memorial University of Newfoundland and a Ph.D. in Cognition and Reading from the University of Alberta. She is the recipient of national and international awards for outstanding research, teaching, and service. She has served on several editorial boards including *Reading Research Quarterly*; and was the senior editor for the *Handbook of Language and Literacy Development: A Roadmap from 0 to 60 Months*. Linda has published numerous books, chapters and articles in top tier language and literacy venues. Her current research includes: the development of a *Test of Early Language and Literacy (TELL)* for children ages 3–8 years; the study of children's reasoning when reading

---

*Author was deceased at the time of publication.

in conventional and dynamic assessment contexts; the exploration of scientific literacy (reading when the content is science); the study of emergent and family literacy; and, the use of fMRI studies in understanding reading development.

Dr. Anat Yarden is Associate Professor and the Head of the Biology Group at the Department of Science Teaching at the Weizmann Institute of Science, Israel. She holds an undergraduate degree in Agricultural Sciences (The Hebrew University of Jerusalem), a master's degree and a Ph.D. in Molecular Biology (Weizmann Institute of Science), and has postdoctoral training in Genetics (Stanford University). The primary theme in all of her academic activities has been the attempt to adapt practices employed by scientists, to the processes by which students and teachers accumulate and advance their knowledge within the discipline of biology. Towards this end, her group pioneered the adaption of primary scientific literature for the teaching and learning of biology in high schools. She is the recipient of a number of prestigious fellowships and awards, including the Fridenberg Foundation Award of the Israel Academy of Sciences and Humanities. She published extensively in leading peer-reviewed journals, is a member of the editorial board of several leading journals in science education, and serves as an associate editor for the Journal of Research in Science Teaching (JRST). She has played active roles in the academic functions of the European Researchers in Didactics of Biology (ERIDOB) and the National Association of Research in Science Teaching (NARST) organizations.

Chapter 1
Prologue: The Origins of the First Adapted Primary Literature

Papers by scientists reporting scientific research have two major advantages as materials for the teaching of science as enquiry. One advantage is obvious. They afford the most authentic, unretouched specimens of enquiry that we can obtain...The second advantage of original papers consists in the richness and relevance of the problems they pose for enquiry into enquiry. (Schwab 1962, p. 73)

The above quotation from Joseph Schwab invites us, reading him 50 years later, to notice scientific inquiry in a different place. We usually are encouraged to find scientific inquiry in the manipulative activities scientists engage to understand the natural world (e.g., National Research Council 2012, p. 24). Schwab tells us we can find inquiry in scientific papers. Indeed, he says those papers are "the most authentic, untouched specimens of inquiry." It is our belief that we can indeed find inquiry in scientific papers, and in reading and the writing of them that underlies and motivates this volume. To think that such papers are the most authentic examples of scientific inquiry truly is an intriguing thought that we ask you to explore with us throughout this book. This book is about what and how scientists read and write and what those practices imply for school science education.

Scientists usually report their research in articles that are published in professional journals. Since the scientists who did the research are also the ones describing it in the articles, such text is termed "primary literature" (Bazerman 1988; Goldman and Bisanz 2002). Primary literature is the way scientists communicate and the way they describe their research and findings. Thus, reading and writing scientific texts are common practices carried out by scientists as part of their scientific research accounting for about 50 % of their working time, and even more for the most productive scientists (Norris and Phillips 2008; Tenopir et al. 2003). Therefore, those practices should play an integral role in the teaching and learning of science alongside other scientific practices such as observing and measuring that already find a central role in science teaching and learning.

At the university and college level, courses that are accompanied by, or even based on reading primary literature, are quite common (Epstein 1970;

© Springer Science+Business Media Dordrecht 2015
A. Yarden et al., *Adapted Primary Literature*, Innovations in Science Education and Technology 22, DOI 10.1007/978-94-017-9759-7_1

Hoskins et al. 2007; Kozeracki et al. 2006). In addition to the possibility to expose these students to contemporary scientific discoveries, the use of primary literature for science learning can also support the development of scientific literacy. Using primary literature for teaching holds the possibility to acquaint students with the rationales for research plans, expose students to research methods and to discussions of their suitability to research questions, acquaint students with the language and structure of scientific communication, develop in students the ability to assess critically the goals and conclusions of scientific research, familiarize students with problems in certain disciplines, and demonstrate the continuity of the scientific research process (Yarden et al. 2001). Thus, teaching with primary literature exposes students to authentic inquiry practices of scientists, as suggested by Norris and Phillips (2008).

Can primary literature also be a way to learn science in high-schools? Learning through primary literature is difficult if not impossible for novices. Yet, Schwab (1962, p. 81) suggested more than 50 years ago that scientific articles "can be edited, excerpted and 'translated'" in order to enable their use for inquiry teaching and learning in high-school. We have taken up this challenge, and developed means to adapt primary literature articles to the knowledge level of high-school students and have termed this educational text genre Adapted Primary Literature (APL, Falk et al. 2008). The adaptation process maintains the canonical structure of the research article as well as its genre, while adapting its contents to the high-school students' prior knowledge level and cognitive abilities (Yarden et al. 2001). We provide several examples of APL articles in part III of the book and in the appendices.

There already exists various means to adapt scientific information to the knowledge level of a target audience. For example, articles in popular scientific magazines (such as Scientific American); science news briefs in daily newspapers (also Journalistic Reported Versions, JRV); popular science programs broadcasted on the public television (like Brainiac); and, perhaps the most ubiquitous of all, textbooks. Each of these communication forms present science in a manner that is very different from the way science is practiced and communicated among scientists. The discourse of science includes not only precise language but also particular ways in which language is used, conclusions are made, ideas are put together, explanations are constructed, and arguments are presented (Krajcik and Sutherland 2010). It thus can come as a surprise to notice that those features usually are absent from the texts and other forms of communication used to relate science to the non-scientific public and particularly to science students in schools.

It is unquestionable that learning to read and write clear and concise expository text, as in summaries of investigations in science class, is an essential preparation for many tasks outside of school (Alberts 2010); and that fundamental literacy practices such as reading, writing and oral discourse are essential to developing scientific literacy (Krajcik and Sutherland 2010). However, school students and educated individuals beyond the school age are known to have difficulties interpreting journalistic reported forms of science as mentioned above (Norris and Phillips 1994; Norris et al. 2003; Phillips and Norris 1999). As well, students in school have problems reading with comprehension scientific textbooks

(Snow 2010). Yet, these various ways of adapting primary scientific literature are meant to assist readers who are not scientists. Our conjecture for explaining this mismatch of intention and outcome is that these various forms of adaptation are insufficiently attuned to the language of science to be able to communicate science in a meaningful way.

By way of contrast, the adapted primary literature genre offers high school students and other non-scientists an opportunity to be exposed to the language of science as it is communicated by scientists. Adapted primary literature exhibits a reduced emphasis on language as a means of transmitting information and an elevated emphasis on language as an interpretive system of sense-making as articulated by Sutton (1998). Intuitively, exposing students to scientific language closer to the way it is used by scientists will present the students with overwhelming difficulties and not provide a solution to difficulties they have reading science. As this book unfolds our joint attempts to realize a vision for the APL genre will be seen to challenge this intuition. The language of science is not the problem. Rather, it is the language that distorts science and exists in tension with science that is problematic.

How the APL Idea Was Initiated

> It is more important that high school students understand how developmental biologists think, than they will be exposed to the wealth of information currently available in the field. (Benny Geiger, personal communication to Anat Yarden, 1996)

At the beginning of 1996, while Anat Yarden was completely immersed in studying the molecular mechanisms involved in cell growth arrest during zebra fish embryonic development, her academic interests took a sharp turn during a meeting with Benny Geiger (Prof. Benjamin Geiger, Department of Molecular Cell Biology, Weizmann Institute of Science). Benny, who served as the Dean of the Feinberg Graduate School at that time, was telling her about plans in the Department of Science Teaching to write a textbook in developmental biology for high school students who chose to major in biology for their matriculation examination (11th and 12th graders). As a scientist in the developmental biology field, Anat thought it would be impossible to cover the basics of the field in the 30 h that were devoted to it as an elective topic in the syllabus. Comprehending those basics requires knowledge of numerous concepts that usually are not familiar to high school students, and she thought that those concepts probably cannot be acquired in the allotted time.

The above quotation from that conversation changed Anat's thoughts about this idea, and actually overturned the way she conceived that biology can be taught and learned. Benny's thoughts are at the basis of our claim above that the language of science is not the problem with students' reading. Benny offered an intriguing suggestion to create a textbook in developmental biology out of scientific papers, in a manner similar to how an issue of a scientific journal, such as Development, is built. The rationale behind this idea was that learning using scientific articles may enable students to acquire knowledge about how scientists conduct their research in the field—the ways they plan their experiments, how they develop their justifications for conducting certain experiments, the means they use to present and discuss the results in the field—while avoiding the emphasis on systematic knowledge acquisition commonly found in developmental biology courses. Since it was clear that high school students are not able to read scientific papers, the proposal was made to adapt to the knowledge level of high school biology majors. These ideas implanted the first seeds of the APL genre, and subsequently led Anat to completely switch fields from a developmental biologist to a researcher in the science education field.

How the APL Idea Was Initially Realized

One of the initial challenges Anat Yarden and her colleague Gilat Brill faced when they started to develop the APL idea was to choose suitable papers for an APL-based curriculum aimed for novices in the field. They approached ten leading developmental biologists and asked them to name breakthrough papers in the field. The rationale was that learning through breakthrough papers might enable high school students to grasp major lines of thinking in developmental biology. At the time, they naively planned to include ten of the resulting papers in the textbook.

At those initial stages of the development of the APL idea, they thought to base an APL paper on data collected from several papers put together in a research paper format especially for an APL-based curriculum. There was even the possibility to carry out experiments and collect data in case the original articles were not clear enough for high school students. The first curriculum in developmental biology included two such papers: (i) Cell migration in the developing embryo: Migration of nuclear crest cells, based on Le Douarin and Teiller (1974) and Le Douarin et al. (1975); and (ii) Regulating head development in the Drosophila embryo: The role of *bicoid* gene, based on Driever and Nusslein-Volhard (1988) and Nusslein-Volhard et al. (1987). The papers displayed a canonical structure of scientific research papers—Introduction, Methods, Results, Discussion (Swales 2001)—as well as the genre in which research papers are usually written—a strong personal voice (Sutton 1998), persuasive and convincing text with high level claims, and hedging (Swales 2001).

A fundamental contribution to the development of the APL idea was made by the Chief Supervisor of Biology Education in Israel at that time, Bruria Agrest. She offered valuable comments and suggestions to multiple revisions of the APL papers in order to adjust them to the knowledge level of high school students. She suggested also to include a very short and basic introduction to the topic. Since scientific articles are written for experts in the field, not all the prior knowledge that is required for comprehending the APL papers can be provided within the papers

Table 1.1 The structure of the first APL-based curriculum: The secrets of embryonic development: study through research (Yarden and Brill 1999)

Section	Description
Part I: Introduction	Basic processes in embryonic development
	Unity among developmental processes
	Fundamental questions in developmental biology
Part II: Collection of research articles	Cell migration in the developing embryo: Migration of nuclear crest cells
	Regulating head development in the Drosophila embryo: The role of *bicoid* gene
	Loss of skeletal muscles in newborn mice bearing a mutation in the *myogenin* gene
	The gene *hedgehog* mediates the polarizing activity in the limb buds

themselves. Such an introduction should help novices become familiar with basic concepts and processes that are required for learning.

The APL-based curriculum was starting to take shape: a brief introduction to the topic accompanied with several APL papers (Table 1.1). The question that remained was about the number of APL papers that can be included in such a curriculum. Eventually, a compromise was made on four papers, instead of the original ten, in order to meet teachers' and supervisors' recommendations on what can be expected from high school students in the 30 h that are allocated for this topic in the syllabus.

In contrast to the approach with the first two adaptations, the next two APL papers were each adapted from a single primary literature paper. Those APL papers were purposely adapted to follow the original paper's logic and use its methodological approach and data. The two papers that were adapted in this manner were: (i) Loss of skeletal muscles in newborn mice bearing a mutation in the *myogenin* gene, based on Hasty et al. (1993) (see a translation to English of this APL paper in Appendix A); and (ii) The gene *hedgehog* mediates the polarizing activity in the limb buds, based on Riddle et al. (1993). The first APL-based curriculum was published and started to be implemented in schools for the first time during 1999 (Yarden and Brill 1999).

How the APL-Based Curriculum Was Initially Implemented

While developing the first version of the APL-based curriculum, Abraham Arcavi asked Anat: "What are they going to do with those articles in class?" The truth is that at that time there was no answer. Needless to say, this question is at the heart of the success of any such novel curriculum. Heda Falk joined the team at that time and started to design ways to implement the developmental biology curriculum in schools (Falk et al. 2003a; Yarden et al. 2001), and to investigate the implementation of the APL-based curriculum in numerous classes (Falk et al. 2005).

Additional studies carried out by Gilat Brill in schools were encouraging. Those studies showed that high school students tended to pose questions that reveal a higher level of inquiry thinking and uniqueness during and following learning using APL in contrast to learning using a textbook (Brill and Yarden 2003). Brill et al. (2004) also found a deeper level of comprehension of an APL text when high school students answered scaffolding questions. Ayelet Baram-Tsabari conducted a study that supported an early hypothesis that learning by using scientific research articles may be a way of developing among students a capacity for scientific ways of thinking among students. Ayelet was able to show that high school biology students who read an APL text better understood the nature of scientific inquiry and raised more scientific criticism of the researchers' work compared to students who read a popular scientific text (Baram-Tsabari and Yarden 2005).

How the APL Idea Was Further Developed

At the beginning of the year 2000 a new syllabus for biological studies in Israel was materializing. Due to the initial success of the APL-based curriculum, it was suggested that a new topic based on APL and termed a Research Topic be included. The objectives of the Research Topic were to: (i) strengthen students' understanding of the Nature of Science (inquiry skills, the history and philosophy of science, quantitative analysis of data, scientific communication); (ii) enable more frequent updates to the syllabus, avoiding the need to change the syllabus itself; (iii) represent the dynamics of biological discoveries by including cutting-edge articles as well as articles that are of utmost importance to the history of biological discoveries; (iv) elicit discussions about socio-scientific and bioethical issues; and (v) encourage teams of teachers to develop a Research Topic themselves as a means for professional development (Israeli Ministry of Education 2003).

Each Research Topic was to be based on knowledge acquired from learning the core topics (Ecology, Systems in the Human Body, and the Living Cell). Each topic comprised an introduction and three APL papers and was intended to be taught for a maximum of 30 h (Israeli Ministry of Education 2003). The three topics that were initially chosen were: (i) Developmental Biology (ii) Biotechnology and (iii) Biodiversity. It was suggested that teachers be able to choose one of the topics. Those three research topics might change, or the APL papers within the three topics might change, from time to time following the biennial advice of the Professional National Committee of Biological Studies in high schools in Israel. Two new APL-based curricula were developed, one in Biotechnology, designed at the Weizmann Institute of Science (Falk et al. 2003b), and a second in Biodiversity designed at the Science Teaching Center of the Hebrew University in Jerusalem (Amir 2005).

Table 1.2 The structure of the APL-based curriculum: Gene tamers – studying biotechnology through research (Falk et al. 2003b)

Section	Description
Part 1: Introduction	Classical vs. modern biotechnology
	Gene cloning in bacteria and in cell cultures
	Genetically modified organisms: plants and animals
	Biotechnology in service of medicine
Part II: Collection of research articles	The development of a biosensor for the detection of genotoxic materials
	Expression of the *Bt* bacterium toxin in chloroplasts of tobacco plants imparts resistance to insects
	Correction of ADA-SCID by bone-marrow stem-cell gene therapy

The structure of the APL-based biotechnology curriculum is identical to the structure of the APL-based developmental biology curriculum (Table 1.2). The introduction to the curriculum lays the groundwork for learning the APL articles by presenting basic concepts and processes in molecular biotechnology. Biotechnology is presented as a practically oriented, problem-solving endeavor. The fact is stressed that many biotechnological solutions, although beneficial, raise new problems. Students are invited to expose the possible drawbacks of present solutions and to suggest theoretical designs for better ones. The main part of the curriculum is composed of three adapted research articles that deal with three different topics: (i) Detection of genotoxic materials in water by bacterial biosensors based on Davidov et al. (2000); (ii) Promotion of plants' resistance to pests by expressing a bacterial toxin based on De Cosa et al. (2001) (see a translation to English of this APL in Appendix B); and (iii) Gene therapy of an immunodeficiency in humans based on Aiuti et al. (2002). These three articles were chosen to represent a variety of organisms used for biotechnological research; and a variety of stages in the biotechnological process, from basic research to field and clinical applications. Topics from cutting-edge research in biotechnology were selected, in the sense that they reached the public media and popular scientific articles have appeared based on them. Other criteria included their adaptability to the APL genre in terms of the presentation of the results, and the compatibility of the scientific background required with high school students' prior knowledge. The adaptation itself was more straightforward this time, as each APL was based on a single Primary Scientific Literature (PSL) paper. Choosing a suitable primary literature paper can be more challenging than the adaptation process itself.

The third APL-based curriculum in biodiversity is similarly structured to include an introduction and three APL papers. However, the APL papers in this curriculum are shorter than the papers in the other two curricula, and they include questions that request students to analyze the presented data instead of providing systematic descriptions of the actual results, as is customary for primary literature papers.

How the APL Acronym Was Coined

Even though the first APL-based curriculum was published in 1999 and the first paper that described the APL approach was published in 2001 (Yarden et al. 2001), the acronym was used for the first time only in 2008 (Falk et al. 2008). The delay was due to a long-lasting debate among the developers of the APL concept about whether the adapted papers can be referred to as primary literature per se or whether those papers represent a different text genre. Indeed all the characteristics of primary literature are represented in the adapted papers. However, there are two major differences between primary literature and APL that eventually convinced the opposers that the APL genre deserves a name of its own: (i) in contrast to primary literature, the writers of the articles are not (necessarily) or not only the scientists who carried out the research; and (ii) the target audience is high school students rather than scientists. The acronym APL was subsequently used in other works (e.g., Norris et al. 2009, 2012; Osborne 2009; Yarden 2009).

How the APL Idea Expanded

During the 2004 annual meeting of the National Association for Research in Science Teaching (NARST) that took place in Vancouver, the three of us met for the first time. Following a presentation on the existence of narrative explanations in primary science sources (Norris et al. 2004) in which we discovered we have mutual interests in the use of scientific texts for science learning in general and primary sources in particular, we decided to meet again to explore those possibilities. This meeting took place during the spring of 2006. At that time, Stephen was able to describe plans to conduct in Canada a study similar to the one reported in Baram-Tsabari and Yarden (2005). His curiosity was centered on whether the effects reported in the 2005 article could be replicated in Canada in a different language, subject, and topic area. The results of that study, since completed, provide confirmatory evidence of the effect of APL on promoting critical thinking (Norris et al. 2012). More recently, Shanahan et al. (2009) has been using successfully a version of APL, which she calls Hybrid Adapted Primary Literature (HAPL), with children in fifth and sixth grades. The 'H' refers to the fact that her adaptations contain a narration at the beginning about the scientists that helps set the personal and social context for the research.

What's Ongoing and Next?

The concept of using APL as part of the curriculum for learning biology in high schools is established in the Israeli syllabus. Two new APL papers have been prepared to replace the first two in the APL-based curriculum in Biotechnology

(Zer-Kavod and Yarden 2013a, b). One of the considerations in choosing the new papers to be adapted is the background knowledge that is provided in the current introductory part to the curriculum, which preferably need not change as frequently as the APL papers. In addition, APL papers for each of the core topics in the syllabus are planned, thus enabling students that did not learn the APL-based curriculum to be exposed to the APL genre. Those papers may enable biology majors (11th–12th graders), who are required to submit an inquiry project as part of their matriculation examination, to be exposed to the APL genre that may serve as a model for the written part of their inquiry project. In addition, APL papers are being prepared for biotechnology majors (11th–12th graders) who are carrying out inquiry projects as part of their matriculation examination. As well, shorter APL papers for the junior-high school population (7th–9th graders) are incorporated in a new curriculum in biology (Ariely and Yarden 2013).

As yet there is no formal recognition of APL in the Canadian science curriculum. Science education is a provincial jurisdiction; it is not certain that if APL were adopted in one province that it would spread to others. At this stage, materials are made available for teachers to use wherever they can find a place within the curriculum. For instance, the introductory grade 12 calculus class in Alberta, contains an optional unit on applications of mathematics to biology. The West Nile Virus module (see Chap. 9) is a perfect choice for teachers wishing to avail of that option.

References

Aiuti, A., Slavin, S., Aker, M., Ficara, F., Deola, S., Mortellaro, A., Morecki, S., Andolfi, G., Tabucchi, A., Carlucci, F., Marinello, E., Cattaneo, F., Vai, S., Servida, P., Miniero, R., Roncarolo, M. G., & Bordignon, C. (2002). Correction of ADA-SCID by stem cell gene therapy combined with nonmyeloablative conditioning. *Science, 296*(5577), 2410–2413. doi:10.1126/science.1070104.

Alberts, B. (2010). Prioritizing science education. *Science, 328*(5977), 405. doi:10.1126/science.1190788.

Amir, R. (2005). *Nature in a world of change: The importance of biodiversity and the causes of its change* (In Hebrew and Arabic, 1st ed.). Jerusalem: The Amos de-Shalit Center for Science Teaching.

Ariely, M., & Yarden, A. (2013). Exploring reproductive systems. In B. Eylon, A. Yarden, & Z. Scherz (Eds.), *Exploring life systems (Grade 8)* (Vol. 2). Rehovot: Department of Science Teaching, Weizmann Institute of Science.

Baram-Tsabari, A., & Yarden, A. (2005). Text genre as a factor in the formation of scientific literacy. *Journal of Research in Science Teaching, 42*(4), 403–428. doi:10.1002/tea.20063.

Bazerman, C. (1988). *Shaping written knowledge: The genre and activity of the experimental article in science*. Madison: The University of Wisconsin Press.

Brill, G., & Yarden, A. (2003). Learning biology through research papers: A stimulus for question-asking by high-school students. *Cell Biology Education, 2*(4), 266–274. doi:10.1187/cbe.02-12-0062.

Brill, G., Falk, H., & Yarden, A. (2004). The learning processes of two high-school biology students when reading primary literature. *International Journal of Science Education, 26*(4), 497–512. doi:10.1080/0950069032000119465.

Davidov, Y., Rosen, R., Smulsky, D. R., Van Dyk, T. K., Vollmer, A. C., Elsemore, D. A., LaRossa, R. A., & Belkin, S. (2000). Improved bacterial SOS promoter: Lux fusions for genotoxicity detection. *Mutation Research, 466*, 97–107. doi:10.1016/S1383-5718(99)00233-8.

De Cosa, B., Moar, W., Lee, S. B., Miller, M., & Daniell, H. (2001). Overexpression of Bt cry2Aa2 operon in chloroplasts leads to formation of insecticidal crystals. *Nature Biotechnology, 19*(1), 71–74.

Driever, W., & Nusslein-Volhard, C. (1988). A gradient of bicoid protein in drosophila embryos. *Cell, 54*, 95–104. doi:10.1016/0092-8674(88)90182-1.

Epstein, H. T. (1970). *A strategy for education*. Oxford: Oxford University Press, Inc.

Falk, H., Brill, G., & Yarden, A. (2003a). *The secrets of embryonic development: Study through research: A teacher's guide* (In Hebrew, 1st ed.). Rehovot: The Amos de-Shalit Center for Science Teaching.

Falk, H., Piontkevitz, Y., Brill, G., Baram, A., & Yarden, A. (2003b). *Gene tamers: Studying biotechnology through research* (In Hebrew and Arabic, 1st ed.). Rehovot: The Amos de-Shalit Center for Science Teaching.

Falk, H., Brill, G., & Yarden, A. (2005). Scaffolding learning through research articles by a multimedia curriculum guide. In M. Ergazaki, J. Lewis, & V. Zogza (Eds.), *Proceedings of the Vth conference of the European researchers in didactics of biology (ERIDOB)* (pp. 175–192). Patra: The University of Patras.

Falk, H., Brill, G., & Yarden, A. (2008). Teaching a biotechnology curriculum based on adapted primary literature. *International Journal of Science Education, 30*, 1841–1866.

Goldman, S. R., & Bisanz, G. L. (2002). Toward a functional analysis of scientific genres: Implications for understanding and learning processes. In J. Otero, J. A. Leon, & A. C. Graesser (Eds.), *The psychology of text comprehension*. Mahwah: Lawrence Erlbaum Associates, Inc.

Hasty, P., Bradley, A., Morris, J. H., Edmondson, D. G., Venuti, J. M., Olson, E. N., & Klein, W. H. (1993). Muscle deficiency and neonatal death in mice with a targeted mutation in the *myogenin* gene. *Nature, 364*, 501–506. doi:10.1038/364501a0.

Hoskins, S. G., Stevens, L. M., & Nehm, R. H. (2007). Selective use of the primary literature transforms the classroom into a virtual laboratory. *Genetics, 176*(3), 1381–1389. doi:10.1534/genetics.107.071183.

Israeli Ministry of Education. (2003). *Syllabus of biological studies (10th–12th grade)*. Jerusalem: State of Israel Ministry of Education Curriculum Center.

Kozeracki, C. A., Carey, M. F., Colicelli, J., & Levis-Fitzgerald, M. (2006). An intensive primary-literature-based teaching program directly benefits undergraduate science majors and facilitates their transition to doctoral programs. *Cell Biology Education, 5*, 340–347. doi:10.1187/cbe.06-02-0144.

Krajcik, J. S., & Sutherland, L. M. (2010). Supporting students in developing literacy in science. *Science, 328*, 456–459. doi:10.1126/science.1182593.

Le Douarin, N. M., & Teiller, M. A. M. (1974). Experimental analysis of the migration and differentiation of neuroblasts of the autonomic nervous system and of neuroectodermal mesenchymal derivatives, using a biological cell marking technique. *Developmental Biology, 41*, 162–184.

Le Douarin, N. M., Teiller, M. A. M., & Le Douarin, G. H. (1975). Cholinergic differentiation of presumptive adrenergic neroblats in interspecific chimeras after heterotopic transplantations. *Proceedings of the National Academy of Sciences U S A, 72*, 728–732.

National Research Council [NRC]. (2012). A framework for K-12 science education: Practices, crosscutting concepts, and core ideas (Committee on a conceptual framework for new K-12 Science Education Standards Board on science education division of behavioral and social sciences and education). Washington, DC: The National Academies Press.

Norris, S. P., & Phillips, L. M. (1994). Interpreting pragmatic meaning when reading popular reports of science. *Journal of Research in Science Teaching, 31*(9), 947–967. doi:10.1002/tea.3660310909.

Norris, S. P., & Phillips, L. M. (2008). Reading as inquiry. In R. A. Duschl & R. E. Grandy (Eds.), *Teaching scientific inquiry: Recommendations for research and implementation* (pp. 233–262). Rotterdam: Sense Publishers.

Norris, S. P., Phillips, L. M., & Korpan, C. A. (2003). University students' interpretation of media reports of science and its relationship to background knowledge, interest, and reading difficulty. *Public Understanding of Science, 12*(2), 123–145. doi:10.1177/09636625030122001.

Norris, S. P., Phillips, L. M., Guilbert, S. M., & Hakimelahi, S. (2004). *Narrative explanations in primary science sources and in science trade books.* Presented at the National Association for Research in Science Teaching Annual International Conference. Vancouver.

Norris, S. P., Macnab, J. S., Wonham, M., & de Vries, G. (2009). West Nile virus: Using adapted primary literature in mathematical biology to teach scientific and mathematical reasoning in high school. *Research in Science Education, 39*(3), 321–329. doi:10.1007/s11165-008-9112-y.

Norris, S. P., Stelnicki, N., & de Vries, G. (2012). Teaching mathematical biology in high school using adapted primary literature. *Research in Science Education, 42*, 633–649.

Nusslein-Volhard, C., Frohnhofer, H. G., & Lehmann, R. (1987). Determination of anteroposterior polarity in drosophila. *Science, 238*, 1675–1681. doi:10.1126/science.3686007.

Osborne, J. (2009). The potential of adapted primary literature (APL) for learning: A response. *Research in Science Education, 39*(3), 397–403. doi:10.1007/s11165-008-9117-6.

Phillips, L. M., & Norris, S. P. (1999). Interpreting popular reports of science: What happens when the reader's world meets the world on paper? *International Journal of Science Education, 21*(3), 317–327. doi:10.1080/095006999290723.

Riddle, R. D., Johnson, R. L., Laufer, E., & Tabin, C. (1993). Sonic hedgehog mediates the polarizing activity of the ZPA. *Cell, 75*, 1401–1416. doi:10.1016/0092-8674(93)90626-2.

Schwab, J. J. (1962). The teaching of science as inquiry. In J. J. Schwab & P. F. Brandwein (Eds.), *The teaching of science.* Cambridge: Harvard University Press.

Shanahan, M. C., Santos, J. S. D., & Morrow, R. (2009). Hybrid adapted primary literature: A strategy to support elementary students in reading about scientific inquiry. *Alberta Science Education Journal, 40*(1), 20–26.

Snow, C. E. (2010). Academic language and the challenge of reading for learning about science. *Science, 328*(5977), 450–452. doi:10.1126/science.1182597.

Sutton, C. (1998). New perspectives on language in science. In B. J. Fraser & K. G. Tobin (Eds.), *International handbook of science education.* Dordrecht: Kluwer.

Swales, J. M. (2001). *Genre analysis: English in academic and research settings.* Cambridge: Cambridge University Press.

Tenopir, C., King, D. W., Boyce, P., Grayson, M., Zhang, Y., & Ebuen, M. (2003). Patterns of journal use by scientists through three evolutionary phases. *D-Lib Magazine, 9*(5).

Yarden, A. (2009). Reading scientific texts: Adapting primary literature for promoting scientific literacy (Guest editorial). *Research in Science Education, 39*(3), 307–311. doi:10.1007/s11165-009-9124-2.

Yarden, A., & Brill, G. (1999). *The secrets of embryonic development: Study through research* (In Hebrew and Arabic, 1st ed.). Rehovot: The Amos de-Shalit Center for Science Teaching.

Yarden, A., Brill, G., & Falk, H. (2001). Primary literature as a basis for a high-school biology curriculum. *Journal of Biological Education, 35*(4), 190–195. doi:10.1080/00219266.2001.9655776.

Zer-Kavod, G., & Yarden, A. (2013a). Engineered bacteria produce biofuel from switchgrass (an adapted primary literature article). *Gene Tamers – Studying Biotechnology Through Research* (In Hebrew, 2nd ed.). Rebovot: Department of Science Teaching, Weizmann Institute of Science.

Zer-Kavod, G., & Yarden, A. (2013b). Immunization – the next generation: Developing genetically engineered eatable plants that can confer immunity against cholera and malaria (an adapted primary literature article). *Gene Tamers – Studying Biotechnology Through Research* (In Hebrew, 2nd ed.). Rehovot: Department of Science Teaching, Weizmann Institute of Science.

Part I
The Theory of Adapted Primary Literature

Chapter 2
Adapting Primary Literature for Promoting Scientific Literacy

This chapter introduces the adapted primary literature (APL) genre. It is a novel genre of science communication that can bring authentic scientific text into the classroom. APL is more closely related to primary scientific literature (PSL), which is used for science communication among scientists, than any other scientific text used in schools (e.g., textbooks, popular scientific text). In response to the current encouragement to incorporate authentic scientific practices into schools we argue that reading APL is such an authentic practice. Since APL provides students with a context that promotes both learning science as inquiry and learning science by inquiry, the question is posed whether APL can advance students' scientific literacy in both the fundamental sense and the derived sense. The chapter deals with three major topics: learning from different genres of scientific texts, the use of authentic scientific text in science and in school, and inquiry that has been conducted on the use of authentic text in teaching scientific literacy. Each of these topics is introduced in this chapter to be explored more fully in later chapters.

Learning from Scientific Texts

Reading scientific text is one of the basic components of scientific literacy. Yet, learning from any text is not a trivial skill and it involves a variety of components. Linking the content of the text with existing knowledge structures, constructing meaning, being able to represent knowledge in different ways, while maintaining motivation and interest are only a few examples. Learning from scientific texts can be even more challenging because those texts often include, in addition to words, diagrams, charts, graphs, images, symbols, and mathematics. Furthermore, learning from scientific texts requires a different purpose of reading than the one required for reading a novel, a poem, or a newspaper. Scientific texts should be read with the aim of extracting information accurately, understanding the arguments, interpreting the meaning, and critically assessing the conclusions.

© Springer Science+Business Media Dordrecht 2015
A. Yarden et al., *Adapted Primary Literature*, Innovations in Science
Education and Technology 22, DOI 10.1007/978-94-017-9759-7_2

Thus, students cannot simply apply reading skills learned elsewhere to scientific texts and expect to have developed effective scientific literacy skills.

Having made these contrasts between reading scientific and other types of texts, we acknowledge that often the differences are ones of degree rather than ones of kind. For example, it is often a goal of newspaper reading to extract information accurately, to grasp arguments, to interpret meaning, and to appraise conclusions. Even novels are informative and often make cases for conclusions. Nevertheless, the mode of presentation of information and argument differs between scientific and non-scientific texts, and readers must decipher this mode of presentation to read effectively.

Reading scientific text in general and PSL in particular is central to scientists' work. PSL is the principal genre of science communication, having been written by the scientists who conducted the research in order to communicate their findings to the scientific community. Reading and critically analyzing PSL is an authentic scientific cognitive activity, grounding scientists' conclusions in the theoretical and empirical work of other scientists. Thus, learning through PSL may provide students unmediated access to a realistic depiction of the nature of science, by exposing them to the authentic inquiry activity of reading scientific texts (Epstein 1970). However, learning through such texts is a near insurmountable challenge for novices, requiring adequate adaptation to be employed as a basis for school and even university curricula (Yarden et al. 2001).

As we discussed in the Prologue, thoughts about the need for such an adaptation led to the development of the APL genre, which is a novel educational text genre that retains many of the characteristics of PSL while adapting its contents to the comprehension level of school students (Falk et al. 2008). Since reading, writing and analyzing primary literature are authentic scientific practices frequently used by scientists, learning through APL can provide an authentic scientific context for learning science. Moreover, because APL presents students with text-based inquiry, it offers an opportunity to promote students' understanding of the nature of science (NOS) and their scientific literacy.

Genres of Scientific Text

Goldman and Bisanz (2002) suggested that there "are three major roles of communication of scientific information in our society. The first is communication among scientists; the second is…popularizing information generated by the scientific community; the third is providing formal education…" (p. 21). These roles serve the needs of different discourse communities, namely scientists; popular scientific writers, and the general public; and textbook and educational authors and their students; respectively. Each discourse community shares a common set of norms for interaction and a language that marks the community and differentiates it from other communities (Swales 2001). A discourse community also shares forms of communication among its members, or genres. A genre is a class of communicative

Table 2.1 Characterizing four genres of scientific texts along six dimensions

Dimension	Genres of scientific text			
	PSL	APL	JRV	Textbooks
Authors	Scientists (Myers 1989; Yore et al. 2004)	Science educators and scientists (Norris et al. 2009; Yarden et al. 2001)	Science journalists (Nwogu 1991)	Science educators and scientists
Target audience	Scientists (Myers 1989; Yore et al. 2004)	Students (high-school science) (Yarden et al. 2001)	General public (Nwogu 1991)	Students (K-12, university)
Main text type	Argumentative (Jimenez-Aleixandre and Federico-Agraso 2009; Suppe 1998)	Argumentative (Norris et al. 2009)	Varying (Expository, Narrative, Argumentative) (Jimenez-Aleixandre and Federico-Agraso 2009; Penney et al. 2003)	Expository (Myers 1992; Norris and Phillips 2008; Penney et al. 2003)
Content	Evidence and reasons to support conclusions (Suppe 1998)	Evidence and reasons to support conclusions (Falk and Yarden 2009; Yarden et al. 2001)	Facts with minimum evidence and reasons (Jimenez-Aleixandre and Federico-Agraso 2009)	Facts (Myers 1992; Newton et al. 2002; Penney et al. 2003)
Organizational structure	Canonical (Suppe 1998; Swales 2001)	Canonical (Baram-Tsabari and Yarden 2005; Yarden et al. 2001)	Non-canonical (Nwogu 1991)	Non-canonical, reflects the knowledge structure of the discipline (Myers 1992)
Presentation of science	Uncertain (Suppe 1998)	Uncertain (Falk and Yarden 2009; Yarden et al. 2001)	Various degrees of certainty (Penney et al. 2003)	Certain (Penney et al. 2003)

Reproduced from Yarden (2009) with minor modifications

events with shared purposes and goals (Goldman and Bisanz 2002). Within the discourse communities mentioned above, PSL represents a genre that is commonly used by the scientific community; Journalistic Reported Versions (JRV), or popular articles published in daily newspapers, can represent a genre commonly used by the general public; and Textbooks can represent a genre commonly used by students in schools, colleges, and universities. Various attributes of these three genres are presented for comparison in Table 2.1, alongside the one which is the focus of this book, namely APL. A comparison of the genres outlined in Table 2.1, which sets APL in the context of the other, better known genres, enables a greater appreciation of the characteristics of APL.

A careful examination of the similarities and differences of the four genres reveals a trend from left to right in the table in the adaptation of primary scientific texts from more to less scientific types of communication. In the far left column, the PSL text represents the genre of communication among scientists. This type of text is written by scientists for scientists. It uses mainly an argumentative text type, includes evidence and reasons to support conclusions (mainly in the Methods and Results sections), is often constructed in a canonical manner (Abstract, Introduction, Methods, Results, Discussion), and presents explicitly the uncertain aspects of science. It should be noted that in some scientific journals (e.g., *Science Magazine*) the canonical structure is less visible to novice readers. Nevertheless, even in those journals all the canonical components appear in the text without specification using headers. It should also be noted that although argumentation is the main text type, PSL can also contain description (e.g., in the Methods and Results sections) and exposition, as in a literature review.

In the far right column, the textbook genre represents the genre of communicating scientific information in the educational system (K-12, college, and university). Textbooks for the K-12 levels are usually authored by science educators and science writers. They are typically written using an expository text type, which reports facts with minimal evidence to support conclusions. Textbooks are frequently structured in order to reflect a presumed knowledge structure of the discipline and present the certain and settled aspects of science. The two extremes, bounded on the one hand by PSL and on the other hand by Textbooks, reflect the contrast between "real science", as it is communicated in each scientific discipline, and "school science", as it is communicated to students in science courses.

In the two columns in the middle of Table 2.1, two different adaptations of PSL are presented, APL and JRV. JRV is somewhat closer to the textbook genre in that it is not structured in a canonical manner, contains mainly facts with minimal evidence to support them, and commonly presents scientific knowledge as certain, although hedging is often present. In contrast, the APL is closer to the PSL according to all of the dimensions examined here, apart from the facts that it is usually not written by the scientists who conducted the research and that its target audience is not scientists but rather science students. (We present two exceptions to this rule in Chaps. 9 and 11. In these cases, scientists involved in the research also assisted in the production of the APL.)

The APL and JRV genres serve different purposes. The JRV genre is important for students as life-long learners, because they are most likely to encounter this genre as an important source of new scientific information following their formal education. APL, on the other hand, is a unique genre developed to provide indirect access to PSL for formal science learning by K-12, college, and university students. It aims to represent science realistically to non-scientists and to promote important aspects of students' scientific literacy that are harder to achieve using textbooks or JRV. It has long been argued that "many science curricula...do not take into account the practical reasoning required in scientific knowledge production" (Norris 1992, p. 196). Considering the various attributes that characterize the four genres of texts discussed above, it is reasonable to assume that the use of argumentative scientific texts, like APL, might be useful in promoting the incorporation of scientific reasoning into the school curriculum. We discuss these possibilities in the subsequent parts of this chapter, as well as in Chap. 4.

Two Perspectives on Scientific Text Genres

We shall now take a deeper look into the various scientific text genres through a closer examination of the attributes that were used for characterizing the various text genres in Table 2.1. The first two dimensions, namely authors and the target audience, enable an examination of the genres from the *sociology of science perspective*, namely from the perspective of the relations between authors and the target audience. The other four dimensions, namely the main text type, content, organizational structure and presentation of science, enable examination of the genres from the *cognitive and linguistic perspectives*; namely, how readers comprehend various text types, the presence or absence of supportive evidence and reasons, the contribution of a canonical or non-canonical text organizational structure, and the presentation of various degrees of certainty of scientific information. Such an examination shows that the APL genre is closer to the JRV and textbooks genres than to the PSL genre in sociological terms, but is closer to the PSL genre in cognitive and linguistic terms.

From the *sociology of science* perspective, the PSL genre is a means of communication among scientists. The discourse community that usually writes and reads the PSL genre comprises scientists that work in the same or closely related scientific domains and investigate similar phenomena or use similar methodologies. Fleck (1979/1935) termed this discourse community "specialized experts", or "the center of the esoteric circle" (Fig. 2.1) of a specific scientific problem. This esoteric circle includes other scientists working on related scientific problems, whom Fleck called "general experts". The richness of each scientific field requires that even within the specialized esoteric circle, a sphere of special experts is distinguished from that of the general ones (Fleck 1979/1935, p. 111).

In contrast to the discourse community of expert scientists, popular science is for non-experts, namely for the large circle of adults, often educated amateurs, that form the exoteric circle. Knowledge presentation for the exoteric circle is characterized by the omission of both detail and controversial opinions, yielding an artificial simplification of scientific knowledge. Therefore, exoteric knowledge can be characterized as simplified, lucid (clear to understanding), and apodictic (absolutely certain) science (Fleck 1979/1935).

As we move away from the esoteric center toward the exoteric periphery, thinking becomes more dominated by an emotive vividness that is integral to the popularization process (Fig. 2.1). The APL genre cannot be placed within the esoteric circle, like the PSL, as it is not aimed to scientists but rather to students of science. However, on the arrow that represents the popularization process of scientific information, APL can be placed within the exoteric circle in close

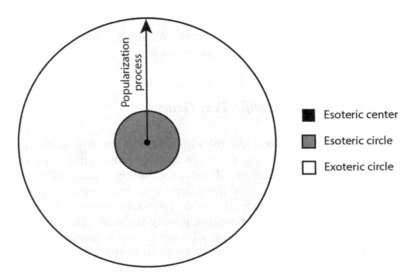

Fig. 2.1 A schematic representation of the distribution of new scientific information. As new scientific information is distributed from the center of the esoteric circle of specialized experts (*Black*) to more general experts (*Dark grey*) and to the exoteric circle of the general public (*Light grey*) it is modified to adjust to the target discourse community

proximity to the esoteric circle. The justification for this placement is that the APL is simplified to the knowledge level of school or early university students and omits unnecessary details (suggesting it belongs in the exoteric circle), yet is not apodictic and does not omit controversial opinions to the same degree as other popular texts like the JRV do. APL is properly placed closer to the esoteric circle than other popular texts, but it certainly belongs within the exoteric circle.

Other interesting features of the different scientific text genres can be discerned when taking the sociology of science perspective. These features include how each genre establishes objectivity, what each regards as constituting fact, and what power relations exist between reader and writer (Parkinson 2001). PSL often achieves objectivity by universalizing propositions through the removal of references to people, time, and place. With the exception of acknowledging authorship, authors of PSL frequently strive to hide their identity and use the "first person plural", namely "we" instead of "I" (Fleck 1979/1935). Thus, PSL functions to persuade readers to accept knowledge claims through a display of objectivity epitomized by hiding human actors behind the forms of the text.

Since PSL undergoes a peer review process before it is published, and is subject to refutation, it is considered highly valuable among members of the scientific community. Moreover, the claims that are put forward in the PSL become facts once they become uncontroversial within the scientific community. Thus, the readers and writers of PSL form a powerful discourse community.

Textbooks also present objectivity by removal of associations with people. However, textbooks take impersonalization even further than PSL, because even authorship of the scientific ideas is not always acknowledged (Parkinson and Adendorff 2004). Moreover, in summarizing the accepted facts of science, textbook writers situate themselves as representatives of the powerful community, which is more powerful in terms of the possession of knowledge than the readers (Parkinson 2001). This relationship between reader and writer is in stark contrast to the situation surrounding the use of PSL, in which readers and writers have much more equal standing in terms of what they know. In addition, in presenting facts that are well accepted by scientists, while omitting the identity of the scientists, textbooks magnify the power differential between the research community and the individual scientist (Myers 1989). In contrast, an interesting feature of JRV, that also is unique to this text genre, is that they are populated with the names of scientists in order to make the text more compelling. The JRV genre usually presents scientists as ordinary rather than as exceptional people of iconic status as in the case in textbooks (Parkinson and Adendorff 2004). APL, on the other hand, strives to objectivity in a manner similar to PSL, by omitting names and places. APL might be considered as being even more impersonal than PSL, because the APL text omits also references to previous works carried out by other scientists. The reason for such impersonalization is that references to PSL usually are not useful for K-12 students.

From the *cognitive and linguistic perspectives* it is interesting to explore how readers comprehend various text genres. The PSL genre is addressed to a scientific audience comprising members knowledgeable in the research area. Therefore, in PSL, knowledge of some terms is assumed and well-known researchers and standard studies are cited (Myers 1989). The text usually is intended to convince other scientists of the strength, power, or truth of the new knowledge claims that are put forward. Thus, referrals to previous prominent scientific publications is often used to support claims (Myers 1989). Moreover, because this text genre serves to report new scientific findings and ideas, it reflects the cutting edge of a scientific field (Goldman and Bisanz 2002). Accordingly, knowledge of the structure and function of each section guides both the writing and the reading of this text genre. Thus, the genre in which PSL is written is argumentative in its nature and represents the argumentative nature of the scientific endeavor as a whole (Suppe 1998). Comprehending the argumentative structure of PSL is not an easy task for novices (e.g., Koeneman et al. 2013). We further discuss the argumentative structure of PSL in Chap. 3, and issues of reading and teaching reading in Chaps. 4 and 6.

In addition to the argumentative structure of the PSL genre, the PSL genre includes numerous scientific and technological terms and it is multi-semiotic in nature. Lemke (2004) claimed that "the language of science is a unique hybrid; natural language as linguistics defines it, extended by the meaning repertoire of mathematics, contextualized by visual representations of many sorts, and embedded in a language (or more properly a 'semiotic') of meaningful, specialized actions afforded by the technological environment in which science is done" (p. 33). At least one, and often more than one, graphical display and at least one mathematical expression per page of running text were reported to be found in typical PSL (Lemke 2004). Thus, the nature of the PSL genre presupposes close and constant integration and cross-contextualization among semiotic modalities. Obviously, interpretation and comprehension of multi-semiotic text such as PSL is not a simple task (Osborne 2002).

As mentioned above, PSL is termed "primary" because it is written by the scientists who conducted the communicated research. In Erwin Schrödinger's words, "A scientist is supposed to have a complete and thorough knowledge, at first hand, of some subjects, and therefore is usually expected not to write on any topic of which he is not a master" (Schrödinger 1944). Accordingly, it has been suggested that PSL is written in language understood by no one outside the authors' fields of expertise (Greene 1997). Indeed, differences between undergraduate students and expert readers of PSL have been identified, and various strategies suggested to enable non-experts to read and comprehend PSL (e.g., Hoskins et al. 2011; van Lacum et al. 2011).

Finally, one of the unique characteristics of PSL that we would like to discuss is the fact that science is presented as uncertain in this scientific text genre. The uncertainty is expressed mainly in hedging, which depicts the tentativeness of the scientist (Varttala 1999) and affords space for readers to anticipate and formulate possible objections to claims (Hyland 1996). Hedged propositions allow writers not only to express their uncertainty concerning the factuality of their statements but also to indicate deference to their readers (Hyland 1994). That is, a hedged text is a type of invitation to the reader to get involved in its assessment, modification, and refinement. This expression of uncertainty in scientific knowledge as it is found in PSL stands in contrast to the certain way in which science often is presented in secondary literature.

Scientific text genres that are written for the general public or students are termed secondary literature. The secondary literature is based on the PSL, and this is why it is termed "secondary". Secondary literature is written (ideally) to match the cognitive level of understanding of its target audiences. Yore and Treagust (2006, p. 296) claimed that "no effective science education program would be complete if it did not support students in acquiring the facility of oral science language and the ability to access, produce, and comprehend the full range of science text and representations."

The three secondary literature genres, textbooks, JRV and APL, each requires adaption to the comprehension level of its target audience. Numerous studies have shown that the adaptations change the language of science from how it is presented in PSL. For example, science textbooks are expository and lack argumentation (Penney et al. 2003), in contrast to PSL which, as pointed out above, and is further elaborated in Chap. 3, is argumentative in nature. The JRV varies between expository, narrative and argumentative (Jimenez-Aleixandre and Federico-Agraso 2009; Penney et al. 2003). As Myers (1992, p. 9) noted "the truly established fact does not need to be stated at all" in PSL, but it does need to be stated in texts written for non-experts, like textbooks or JRV, in order to make them comprehensible to the readers. Thus, it seems that the process that makes such text genres comprehensible to their target audiences also renders them less argumentative in nature, further distancing them from the language of science as it is communicated among the scientific community. Moreover, photographs, diagrams, graphs, and other illustrations serve different functions in the various text genres. Illustrations in textbooks are often not included to provide evidence to support conclusions, but rather to decorate the text or to serve specific pedagogical purposes (Myers 1992). Similarly, illustrations in JRV articles often serve to decorate or attract readers' attention, rather than to elaborate concepts or to provide evidence in support of claims made in the text.

The uncertainty manifested in PSL reflects the true status of the scientific information, as "they [really valuable experiments] are all of them uncertain, incomplete, and unique. And when experiments become certain, precise, and reproducible at any time, they no longer are necessary for research purposes proper but function only for demonstration or ad hoc determinations." (Fleck 1979/1935, p. 85). It is therefore startling how almost all statements in science textbooks are presented as being true when the textbooks are supposed to portray science. In addition, the range of meta-scientific language use in textbooks is largely limited to observational words, words that describe the process of doing research, and relational words, such as 'cause' (Penney et al. 2003). In JRV the situation is less severe in this sense, as only some of the statements are presented as true and meta-scientific language is somewhat used (Penney et al. 2003).

The characteristics of secondary literature that make it depart from PSL in meaning and form are not inevitable. APL is designed with the purpose of keeping it as similar as possible to PSL, while at the same time making it comprehensible to the target audience (Yarden et al. 2001). Thus, similarly to PSL, APL is argumentative and the visualizations that are included usually are taken directly from the PSL source, or modified and adjusted to the comprehension level of the target audience. Modifications might include, for example omissions of specific illustrations or parts of illustrations to adjust to the APL text (Yarden et al. 2001) or additions to illustrations to make them easier to interpret for students.

Osborne (2002) claimed that "science cannot be understood without an exploration of its language" (p. 212), while referring to reading, writing, and argument as central to any conception of science. We claim that APL represents the argumentative structure, scientific content, organizational structure, and uncertainty that are found in PSL. As such, APL exhibits the language of science that is often absent from the other text genres used for science learning. Thus, APL enables authentic scientific texts to be brought to a wide audience outside of the scientific community, namely, to those in the exoteric circle (Fig. 2.1). In this book we discuss how this wider dissemination is realized successfully.

Authentic Scientific Texts in Science and in School

The recent framework for K-12 science education (National Research Council [NRC] 2012) emphasizes the importance of developing students' scientific knowledge while strengthening their competency with related practices. It is argued that engaging in the practices of science helps students understand how scientific knowledge develops; and that such a direct involvement gives them an appreciation of the wide range of approaches that are used to investigate, model, and explain the world.

Attempts to make science learning better resemble authentic scientific practices have led to educational reforms at least since Dewey (1964). Based on the notion that knowledge is situated, being in part a product of the activity, context, and culture in which it is developed and used, Brown et al. (1989) suggested that authentic activities are the "ordinary practices of the culture" (p. 34). They argued that authentic activity is important for learners "because it is the only way they gain access to the standpoint that enables practitioners to act meaningfully and purposefully" (p. 36). Numerous educational researchers have adopted authentic scientific practices, as they are practiced by the scientific community, as a basis for the teaching and learning of science (e.g., Edelson 1998). Those approaches have been termed for example "scientists' science" (Rahm et al. 2003), "doing science", (Jimenez-Aleixandre et al. 2000), "science practice" (Edelson 1998), or "real-world science" (Lee and Songer 2003). Buxton (2006) coined this perspective of authenticity as canonical since it is aligned with both the Western scientific canon and with the canon for science education standards in the US (NRC 2007), Europe (European Union 2006), and elsewhere. It should be noted that we will not refer here to other perspectives on authenticity that are relevant to the teaching and learning of science, namely, the youth-centered perspective and the contextual perspective (Buxton 2006), and the real-world authenticity and authentic assessment approach (Shaffer and Resnick 1999), as they are less relevant when discussing reading in science.

Chinn and Malhotra (2002) argued that scientific inquiry tasks given to students in school usually reflect science as "simple, certain, algorithmic and focused at a surface level of observation" (p. 190). Accordingly, the cognitive processes needed to succeed in many school tasks are often qualitatively different or even antithetical to the cognitive processes and epistemology needed to engage in real scientific research. At the level of classroom practice, the students' classroom responsibilities, roles, and routines usually differ from scientific practices (Ford and Wargo 2007). Similarly, the text genres used for the teaching and learning of science are significantly different from those used by scientists. In contrast to scientists who extensively study other scientists' research using the various authentic scientific texts mentioned above, reading expert research reports plays almost no role at all in simple forms of school science. At most, students conduct their own research and prepare scientific reports; but even then, students do not study a body of research that has passed review by experts in the field, as is customary in authentic scientific research (Chinn and Malhotra 2002). Thus, the APL genre, which retains the characteristics of PSL while adapting its contents to the comprehension level of school students, can be used for promoting the canonical perspective of authentic science in schools. In the subsequent section we discuss the use of texts in authentic scientific research and the means to adapt such practices into the teaching and learning of science.

The Use of Texts in Authentic Scientific Research

As discussed above, a prominent feature of scientists' research practices is communicating scientific information among the esoteric circle of experts. Scientists spend a significant portion of their time in reading and writing scientific texts (Tenopir and King 2001), and those practices are constitutive parts of authentic science (Norris and Phillips 2003, 2008). In addition to reading and writing such texts, scientists are continuously engaged in reviewing, evaluating, and critiquing texts that are written by other scientists. The texts that are commonly used by scientists include PSL, as well as research presentations, grant proposals, theses and dissertations, review articles and the like, that altogether represent the discourse within a community of experts that is investigating a specific scientific matter. The use of such authentic scientific texts advances scientific knowledge and skills among the scientific community at large. The scientific endeavor is characterized by attitudes of uncertainty as both the techniques and results of scientific inquiry are subject to continual re-examination. Within the processes of coordination and re-examination, an important role is played by argumentation.

Argumentation plays an essential role in the processes of writing, reading, and making sense of the findings of others (Bazerman 1988). The discursive process employed by scientists to construct concepts and theories from empirical data is actually argumentation that is used as a means of presenting and weighing evidence, assessing alternatives, interpreting texts and evaluating a concept's potential (Latour and Woolgar 1986). Thus, authentic scientific texts like PSL serve as arguments for the scientists' claims (Bazerman 1988). The information that is provided in those authentic texts is often new knowledge that either supports or contradicts existing theories or facts. Suppe (1998) argues that in addition to the text itself every paragraph and illustration contributes to establishing the claims in a scientific paper. Indeed, one of the prominent characteristics of authentic scientific texts is the unique way in which scientific information is presented in them and mainly the uncertainty of the scientific information and the dynamic nature of scientific knowledge. Authentic scientific texts, like PSL, usually suggest a model and make a sustained case for its explanatory credentials, but they also aim to ensure that data conflicting with the hypothesis being advanced do not discredit it. Thus, the data presented in scientific papers may not be always in line with the hypothesis that is suggested in the same scientific articles or in other articles (Lipton 1998), thus exemplifying other uncertain aspects that are expressed in PSL.

In the dynamic process of allowing new knowledge claims into the scientific canon through authentic means of science communication (e.g., through PSL), conclusions should cohere with the range of evidence, even when data are conflicting, demanding high scientific standards to form consistency between theoretical models and observations of phenomena (Chinn and Malhotra 2002; Hogan and Maglienti 2001). Indeed, authentic scientific research is a complex activity employing complex theories, highly specialized knowledge, and elaborate and expensive procedures. This complexity involves reasoning processes that are reflected by practices that include, among others: transformation of observations, coordination of theoretical models with multiple sets of contradictory data, and rationally and regularly discounting anomalous data. This complexity calls for reasoning processes that are qualitatively different from those required by simple school tasks and is rarely reflected in school science (Chinn and Malhotra 2002).

The Use of Authentic Texts in School Science

As discussed above, school science differs dramatically from authentic science that is carried out by scientists (Chinn and Malhotra 2002). Indeed, students' disciplinary content knowledge, their motivational goals, commitment, resources and social organization, are only a few characteristics that are dramatically different from those of scientists. Thus, special adaptations are required in order to bring science into schools in as authentic a form as possible. Understanding the nature and also the source of differences in how scientists and nonscientists think can inform the adaptation efforts (Hogan and Maglienti 2001). Past calls for promoting the acquisition of inquiry skills using hands-on activities are now being replaced by calls to promote the rhetoric of science based on criticism and argumentation when seeking evidence and reasons for the ideas or knowledge claims (i.e., Jimenez-Aleixandre et al. 2000). Instead of viewing practical experimental work as the basis of students' scientific procedural practices, it is suggested that it be valued for the role it plays in providing evidence for knowledge claims (Millar and Osborne 1998).

Two types of educational interventions that can present an authentic context in schools, namely first-hand and second-hand investigations, have been suggested (Palincsar and Magnusson 2000). First-hand investigations allow students to solve problems themselves, via hands-on projects or laboratory work, namely via the traditional means of teaching and learning of science through inquiry. Second-hand investigations present students with results that have been obtained by scientists, through software resources or texts (i.e., Hapgood et al. 2004). The authenticity of these investigations is enforced by the fact that much of what scientists know has been acquired through the thoughts and experiences of others—that is, through learning in a second-hand manner. Even though one of the most prominent scientific practices is communicating scientific information, namely reading and writing scientific information, there is a strong apprehension about the use of texts in school science, particularly in the inquiry science tradition (Cervetti et al. 2006). Therefore, text other than that found in textbooks has not typically been prominent in the context of inquiry science curriculum and pedagogy (Palincsar and Magnusson 2000). Considering the characteristics of text genres that are commonly used in schools, which are mainly textbooks, this is not surprising as textbooks usually do not enable authentic science practices. The textbook genre, which usually presents science as a collection of certain facts fails (Table 2.1), to present the argumentative nature and uncertainty aspects of science. We argue that the APL genre, which includes evidence and reasons to support conclusions and presents the uncertain aspects of science (Table 2.1), similarly to the PSL genre, enables authentic science including the authentic scientific practice of reading scientific text. Thus, in contrast to most common textbooks, APL provides a means to bring second-hand investigations into the classroom.

Two main approaches were identified by Radinsky et al. (2001) for designing authentic curricula that allow enculturation of students into the 'ways of knowing' that are commonly used in the specific scientific discipline: 'simulation' and 'participation'. The 'simulation' approach involves creating an imitation of a professional practice within the context of the classroom, by designing materials, tools, assignments, and interactions that are in line with the activities of the professional community. The 'participation' approach involves creating opportunities for students to take part in the actual work of a professional scientific community, thus allowing them to learn about elements of the practice that may not be captured in a simulation. The APL genre provides both a simulation and an opportunity for participation in a scientific practice in the classroom, thus enabling enculturation of students into an authentic scientific experience.

Inquiry into Authentic Scientific Texts and the Promotion of Scientific Literacy

Norris and Phillips (2003) suggested two distinct meanings of scientific literacy: the fundamental sense—the ability to read, interpret and write a scientific text—and the derived sense, which is the knowledge of scientific ideas and the ability to use them in a scientific manner. These two meanings of scientific literacy are not independent but dialectic, as they support and complement each other (Norris and Phillips 2003).

The emerging idea is that the fundamental sense is critical in supporting the derived sense. The opposite is also true: "How, for instance, can we imagine interpreting a science text by making interconnections throughout the text (an activity we have associated with the fundamental sense) without being knowledgeable of the substantive content of science (literate in the derived sense)?" (Norris and Phillips 2003, p. 236). One of the aspects of the derived meaning of scientific literacy is the ability to think scientifically (DeBoer 2000). Since, from an epistemological perspective, inquiry is simply the process of doing science (Schwab 1962), we regard inquiry thinking as one aspect of the derived sense of scientific literacy.

Inquiry learning has been classified as learning science as inquiry and by inquiry (Tamir 1985). Learning science as inquiry includes learning about the way in which the scientific endeavor progresses, and analyzing the inquiry process performed by others, sometimes using historical examples (Bybee 2000; Schwab 1962). This characterization converges with the understanding of the Nature of Science (NOS), which refers to the values, influences and limitations intrinsic to scientific knowledge and to science as a human endeavor (Schwartz et al. 2004). Aspects of the

NOS that are considered important objectives of science education include: understanding the nature, production, and validation of scientific knowledge; the internal and external sociology of science, and the processes of science (Aikenhead and Ryan 1992). Learning by inquiry, or learning "the abilities necessary to do scientific inquiry" (Bybee 2000), involves the learner in raising research questions, generating a hypotheses, designing experiments to verify them, constructing and analyzing evidence-based arguments, recognizing alternative explanations, and communicating scientific arguments.

Schwab suggested 'enquiry into enquiry' as an optimal approach to promoting inquiry learning, claiming that: "The complete enquiring classroom would have two aspects. On the one hand, it would exhibit science as enquiry. On the other hand, the student would be led to enquire into these materials" (Schwab 1962, p. 65). Schwab also claimed that translated and excerpted primary literature can be an optimal intervention for promoting 'enquiry into enquiry'. Since APL provides students with a context that promotes both dimensions of inquiry learning (Falk et al. 2008), an important theoretical question is whether it can also provide a context for promoting both senses of scientific literacy. The claim that text and inquiry require and promote similar cognitive processes (Gaskins et al. 1994) might support this theoretical framework. If suitable scientific texts can promote both inquiry and literacy, a further question would be: in what way do these texts relate to the existing gap between school science and authentic science?

References

Aikenhead, G. S., & Ryan, A. G. (1992). The development of a new instrument: "Views on Science-Technology-Society" (VOSTS). *Science Education, 76*(5), 477–491.

Baram-Tsabari, A., & Yarden, A. (2005). Text genre as a factor in the formation of scientific literacy. *Journal of Research in Science Teaching, 42*(4), 403–428.

Bazerman, C. (1988). *Shaping written knowledge: The genre and activity of the experimental article in science*. Madison: The University of Wisconsin Press.

Brown, J. S., Collins, A., & Duguid, P. (1989). Situated cognition and the culture of learning. *Educational Researcher, 18*(1), 32–42.

Buxton, C. A. (2006). Creating contextually authentic science in a "low performing" urban elementary school. *Journal of Research in Science Teaching, 43*(7), 695–721.

Bybee, R. W. (2000). Teaching science as inquiry. In J. Minstrell & E. H. van Zee (Eds.), *Inquiring into inquiry learning and teaching in science* (pp. 20–62). Washington, DC: American Association for the Advancement of Science.

Cervetti, G. N., Pearson, P. D., Bravo, M. A., & Barber, J. (2006). Reading and writing in the service of inquiry-based science. In R. Douglas, M. Klentschy, & K. Worth (Eds.), *Linking science and literacy in the K-8 classroom*. Arlington: National Science Teacher Association.

Chinn, C. A., & Malhotra, B. A. (2002). Epistemologically authentic inquiry in schools: A theoretical framework for evaluating inquiry tasks. *Science Education, 86*, 175–218.

DeBoer, G. E. (2000). Scientific literacy: Another look at its historical and contemporary meanings and its relationship to science education reform. *Journal of Research in Science Teaching, 37*(6), 582–601.

Dewey, J. (1964). Science as subject matter and as method. In R. D. Archambault (Ed.), *John Dewey on education: Selected writings* (pp. 121–127). Chicago: University of Chicago Press.

Edelson, D. C. (1998). Realising authentic science learning through the adaptation of scientific practice. In B. J. Fraser & K. G. Tobin (Eds.), *International handbook of science education* (pp. 317–331). Dordrecht: Kluwer.

Epstein, H. T. (1970). *A strategy for education.* Oxford: Oxford University Press, Inc.

European Union. (2006). Recommendation of the European Parliament and of the Council of 18 December 2006 on key competences for lifelong learning. *Official Journal of the European Union*, 30-12-2006, L 394/310–L 394/318.

Falk, H., & Yarden, A. (2009). "Here the scientists explain what I said." Coordination practices elicited during the enactment of the results and discussion sections of adapted primary literature. *Research in Science Education, 39*(3), 349–383.

Falk, H., Brill, G., & Yarden, A. (2008). Teaching a biotechnology curriculum based on adapted primary literature. *International Journal of Science Education, 30*(14), 1841–1866.

Fleck, L. (1979/1935). *Genesis and development of a scientific fact* (F. Bradley & T. J. Trenn, Trans.). Chicago: The University of Chicago Press.

Ford, M. J., & Wargo, B. M. (2007). Routines, roles, and responsibilities for aligning scientific and classroom practices. *Science Education, 91*, 133–157.

Gaskins, I. W., Guthrie, J. T., Satlow, E., Ostertag, J., Six, L., Byrne, J., & Connor, B. (1994). Integrating instruction of science, reading, and writing: Goals, teacher development, and assessment. *Journal of Research in Science Teaching, 31*(9), 1039–1056.

Goldman, S. R., & Bisanz, G. L. (2002). Toward a functional analysis of scientific genres: Implications for understanding and learning processes. In J. Otero, J. A. Leon, & A. C. Graesser (Eds.), *The psychology of text comprehension.* Mahwah: Lawrence Erlbaum Associates Publication.

Greene, M. T. (1997). What cannot be said in science. *Nature, 388*(6643), 619–620.

Hapgood, S., Magnusson, S. J., & Palincsar, A. S. (2004). Teacher, text, and experience: A case of young children's scientific inquiry. *Journal of the Learning of Sciences, 13*(4), 455–505.

Hogan, K., & Maglienti, M. (2001). Comparing the epistemological underpinnings of students' and scientists' reasoning about conclusions. *Journal of Research in Science Teaching, 36*(6), 663–687.

Hoskins, S. G., Lopatto, D., & Stevens, L. M. (2011). The C.R.E.A.T.E. approach to primary literature shifts undergraduates' self-assessed ability to read and analyze journal articles, attitudes about science, and epistemological beliefs. *CBE Life Sciences Education, 10*(4), 368–378.

Hyland, K. (1994). Hedging in academic writing and EAF textbooks. *English for Specific Purposes, 13*(3), 239–256.

Hyland, K. (1996). Writing without conviction? Hedging in science research articles. *Applied Linguistics, 17*(4), 433–454.

Jimenez-Aleixandre, M. P., & Federico-Agraso, M. (2009). Justification and persuasion about cloning: Arguments in Hwang's paper and journalistic reported versions. *Research in Science Education, 39*(3), 331–347.

Jimenez-Aleixandre, M. P., Bugallo Rodriguez, A., & Duschl, R. A. (2000). "Doing the lesson" or "doing science": Argument in high school genetics. *Science Education, 84*, 757–792.

Koeneman, M., Goedhart, M., & Ossevoort, M. (2013). Introducing pre-university students to primary scientific literature through argumentation analysis. *Research in Science Education, 43*(5), 2009–2034. doi:10.1007/s11165-012-9341-y.

Latour, B., & Woolgar, S. (1986). *Laboratory life: The construction of scientific facts.* Princeton: Princeton University Press.

Lee, H. S., & Songer, N. B. (2003). Making authentic science accessible to students. *International Journal of Science Education, 25*(8), 923–948.

Lemke, J. L. (2004). The literacies of science. In E. W. Saul (Ed.), *Crossing borders in literacy and science instruction* (pp. 33–47). Newark/Arlington: International Reading Association/ National Science Teachers Association.

Lipton, P. (1998). The best explanation of a scientific paper. *Philosophy of Science, 65*(3), 406–410.

Millar, R., & Osborne, J. (1998). *Beyond 2000: Science education for the future*. London: King's College.

Myers, G. (1989). The pragmatics of politeness in scientific articles. *Applied Linguistics, 10*(1), 1–35.

Myers, G. A. (1992). Textbooks and the sociology of scientific knowledge. *English for Specific Purposes, 11*, 3–17.

National Research Council [NRC]. (2007). *Taking science to school: Learning and teaching science in grades K-8*. Washington, DC: The National Academies Press.

National Research Council [NRC]. (2012). *A framework for K-12 science education: Practices, crosscutting concepts, and core ideas*. Washington, DC: The National Academies Press.

Newton, L. D., Newton, D. P., Blake, A., & Brown, K. (2002). Do primary school science books for children show a concern for explanatory understanding? *Research in Science and Technological Education, 20*(2), 227–240.

Norris, S. P. (1992). Practical reasoning in the production of scientific knowledge. In R. A. Duschl & R. J. Hamilton (Eds.), *Philosophy of science, cognitive psychology, and educational theory and practice* (pp. 195–225). Albany: State University of New York Press.

Norris, S. P., & Phillips, L. M. (2003). How literacy in its fundamental sense is central to scientific literacy. *Science Education, 87*, 224–240.

Norris, S. P., & Phillips, L. M. (2008). Reading as inquiry. In R. A. Duschl & R. E. Grandy (Eds.), *Teaching scientific inquiry: Recommendations for research and implementation* (pp. 233–262). Rotterdam: Sense Publishers.

Norris, S. P., Macnab, J. S., Wonham, M., & de Vries, G. (2009). West Nile virus: Using adapted primary literature in mathematical biology to teach scientific and mathematical reasoning in high school. *Research in Science Education, 39*(3), 321–329.

Nwogu, K. N. (1991). Structure of science popularizations: A genre-analysis approach to the schema of popularized medical texts. *English for Specific Purpose, 10*, 111–123.

Osborne, J. (2002). Science without literacy: A ship without a sail. *Cambridge Journal of Education, 32*(2), 203–218.

Palincsar, A. S., & Magnusson, S. J. (2000). *The interplay of firsthand and text-based investigations in science education (CIERA REPORT #2-007)*. Ann Arbor: University of Michigan.

Parkinson, J. (2001). *Popular and academic genres of science: A comparison, with suggestions for pedagogical applications*. Durban: University of Natal.

Parkinson, J., & Adendorff, R. (2004). The use of popular science articles in teaching scientific literacy. *English for Specific Purposes, 23*, 379–396.

Penney, K., Norris, S. P., Phillips, L. M., & Clark, G. (2003). The anatomy of junior high school science textbooks: An analysis of textual characteristics and a comparison to media reports of science. *Canadian Journal of Science, Mathematics, and Technology Education, 3*(4), 415–436.

Radinsky, J., Bouillion, L., Lento, E. M., & Gomez, L. M. (2001). Mutual benefit partnership: A curricular design for authenticity. *Journal of Curriculum Studies, 33*(4), 405–430.

Rahm, J., Miller, H. C., Hartley, L., & Moore, J. C. (2003). The value of an emergent notion of authenticity: Examples from two student/teacher-scientist partnership programs. *Journal of Research in Science Teaching, 40*(8), 737–756.

Schrödinger, E. (1944). *What is life?* Cambridge: Cambridge University Press.

Schwab, J. J. (1962). The teaching of science as enquiry. In J. J. Schwab & P. F. Brandwein (Eds.), *The teaching of science*. Cambridge: Harvard University Press.

Schwartz, R. S., Lederman, N. G., & Crawford, B. A. (2004). Developing views of nature of science in an authentic context: An explicit approach to bridging the gap between nature of science and scientific inquiry. *Science Education, 88*, 610–645.

Shaffer, D. W., & Resnick, M. (1999). "Thick" authenticity: New media and authentic learning. *Journal of Interactive Learning Research, 10*(2), 195–215.

Suppe, F. (1998). The structure of a scientific paper. *Philosophy of Science, 65*, 381–405.

Swales, J. M. (2001). *Genre analysis: English in academic and research settings* (First edition 1990 ed.). Cambridge: Cambridge University Press.

Tamir, P. (1985). Content analysis focusing on inquiry. *Journal of Curriculum Studies, 17*(1), 87–94.

Tenopir, C., & King, D. W. (2001). The use and value of scientific journals: Past, present and future. *Serials, 14*(2), 113–120.

van Lacum, E., Ossevoort, M., Buikema, H., & Goedhart, M. (2011). First experiences with reading primary literature by undergraduate life science students. *International Journal of Science Education, 34*(12), 1795–1821.

Varttala, T. (1999). Remarks on the communicative functions of Hedging in popular scientific and specialist research articles on medicine. *English for Specific Purposes, 18*(2), 177–200.

Yarden, A. (2009). Guest editorial – Reading scientific texts: Adapting primary literature for promoting scientific literacy. *Research in Science Education, 39*(3), 307–311.

Yarden, A., Brill, G., & Falk, H. (2001). Primary literature as a basis for a high-school biology curriculum. *Journal of Biological Education, 35*(4), 190–195.

Yore, L. D., Hand, B., Goldman, S. R., Hildebrand, G. M., Osborne, J. F., Treagust, D. F., & Wallace, C. S. (2004). New directions in language and science education research. *Reading Research Quarterly, 39*(3), 347–352.

Yore, L. D., & Treagust, D. F. (2006). Current realities and future possibilities: Language and science literacy—empowering research and informing instruction. *International Journal of Science Education, 28*(2–3), 291–314.

Chapter 3
Foundations for Conceptualizing APL

This chapter opens with a discussion on the importance of reading to science and to scientists. The claim is made that the empirical basis of science is well-canvassed in the school curriculum but that the basis of science in literacy is almost totally neglected. We explore the implications of this situation for students' science learning and understanding. The remainder of the chapter deals in one way or another with the structure of scientific text. First, we delineate three ways in which the structure of text can be understood and illustrate these distinctions by drawing upon published scientific work. Second, we examine how the structure of scientific texts varies within and between scientific disciplines. The upshot of the analysis is that structure varies considerably both within and between disciplines and that perceived patterns likely have more to do with the overall purpose of the research and with scientists' personal styles and communicative preferences. Third, we examine the epistemology underlying scientific texts and conclude, using examples to illustrate, that scientific writings are underlain by fallible rationality. Finally, we look at meta-scientific language that is found in scientific texts, and show how scientists use such language to, in a sense, provide the perspective of an outsider so that they can deal critically with their own work.

Reading Is Important to Science and Scientists

The Empirical Basis of Science

A constitutive feature of science is its empirical basis. Scientific ideas are designed to describe and explain what happens in the world, and scientists test these ideas by collecting data to determine the degree to which they fit the world. We might disagree over when an idea explains and how we use data to judge the fit between ideas and the world, and even disagree over what "fit" means. Nevertheless, we

© Springer Science+Business Media Dordrecht 2015
A. Yarden et al., *Adapted Primary Literature*, Innovations in Science Education and Technology 22, DOI 10.1007/978-94-017-9759-7_3

would agree that no enterprise properly could be called "science" whose ideas were not about the make-up of the world and that did not evaluate those ideas for accuracy by comparing them to the world. It is this necessity for an empirical basis that prompts us to call it "a constitutive feature of science"—without an empirical basis, science as we know it would not exist.

Typically, science curricula have set out to teach students that science has this empirical characteristic, and many students come away with this idea, even if they misunderstand science in other ways. Thus, it is common for school students, when asked to describe science and what scientists do, to mention that scientists study the world, do experiments, and test things out. Mead and Métraux (1957) documented these tendencies more than five decades ago among a sample of nearly 35,000 American high school students, who were asked to write brief essays to complete statements such as the following: "When I think about a scientist, I think of" (1957, p. 385). They provided a composite image of a scientist from responses made by students:

> The scientist is a man who wears a white coat and works in a laboratory. He is elderly or middle aged and wears glasses. . . . He is surrounded by equipment . . . The sparkling white laboratory is full of sounds. . . . He spends his days doing experiments . . . (pp. 386–387)

When asked to draw scientists, students often draw them surrounded by test tubes, weigh scales, and other instruments. Using his now widely known and used, the Draw-a-Scientist Test, Chambers (1983), writing more than 25 years after Mead and Métraux and studying students in kindergarten to grade five from Australia, Canada, and the United States, reported that the stereotypic image of the scientist, which Mead and Metraux [sic] examined in high school students, was also found to appear among students at the grade school level. Indeed, multiple studies over the years of different ages, ethnicities, and nationalities of students have reported coincident findings (e.g., Finson 2003; Fung 2002; Mason et al. 1991; Song and Kim 1999) Stereotypical images of science often are revealed in what they say, write, and draw. Nevertheless, in their descriptions and images there often is a kernel of truth about science's empirical nature.

Science curricula typically contain some sort of regular hands-on experience, laboratory or practical work, which both creates and reinforces this empirical view of science in students' minds. Of course, their conceptions of the nature of experiments, of a scientific test, of evidence, and of the workings and functions of scientific instruments might be badly distorted. Yet, they seem to learn quite early that science is empirical. It may well seem that this is not much of an accomplishment, but it appears to be an enduring lesson and one upon which science education profitably can build.

The Literacy Basis of Science

We have argued (Norris and Phillips 2003; Norris et al. 2009) that having a basis in literacy is another constitutive feature of science. It is impossible to imagine science as we know it without literacy. To be specific, science, as distinct from individual pieces of knowledge, could not arise or be conducted in an oral context, despite the fact that oral exchanges are crucial to the development and continuation of science. There are both psychological and practical necessities that demand literacy be at the basis of science. Memory is painfully finite and unreliable; having scientific ideas written down so that they can be consulted helps mitigate the shortcomings of memory. Moreover, science is more than knowledge—it is also a corpus of practices for building that knowledge. We cannot imagine, to take one such practice, an effective exchange of scientific ideas and peer review without documents that can be sent from one scientist to another. However, more fundamental than any psychological or practical reason for the necessity of literacy to science are epistemological reasons. Without literacy and the recording of scientific ideas using print, there is no means of making scientific ideas stand still, metaphorically thinking, so that they can be assessed. When a written scientific document is created, it assumes a relative fixity in two senses: first, in physical form and presentation (the actual words, sentences, punctuation, tables of data, graphs, reference lists, and so on); and, second, in literal meaning.

The first sense of fixity it easy to grasp, so we shall say no more about it. The fixity of literal meaning is somewhat more subtle. Take for instance a table of data that relates the quantity pressure to the quantity temperature. Scientists might argue over the interpretation of these data. Possibly, there is more than one interpretation that makes sense. However, it is not possible for every interpretation to make sense because the literal meaning of the data constrains the space of viable interpretations. Within science it is not legitimate to interpret the data point (2.0 kPa, 275.6 K) as meaning (3.0 kPa, 278.6 K) without specifying some error in the data collection or tabulation to justify this interpretation. The statement seems so obvious as not to be worth saying. However, literal meaning is based upon assumptions so basic that they hardly ever percolate to the surface of consciousness. One of those assumptions is that without such constraints on legitimacy as the one just mentioned, interpretation is not possible, because without constraint any interpretation is as good as any other. If any interpretation is as good as any other, evaluation is not possible, which brings us back to the epistemological implications of literacy. It is literacy that makes possible the evaluation of scientific ideas because literacy makes it possible to fix literal meaning in a permanent record.

The point is worth exploring further, because literal meaning is only relatively fixed in science, that is, fixed compared to conclusions based on that meaning. It is less than two decades ago since a special issue of the *Journal of Research in Science Teaching* (1994, Volume 31, Issue 9) provided recognition for the central and important role of literacy in science. Before that time, and even in much research since, all properly scientific meanings were assumed to be literal. This point might make immediate intuitive sense—after all, we normally do not think of scientific language being figurative, which often is contrasted to the literal. Figurative language finds its use in poetry and other literary forms. Thus, a reported scientific observation should have a universal meaning, that is, a meaning independent of the context of its use and of the scientists reading the report. Normally, scientists do not

search for the metaphors behind observation reports. The switch in science education research to a focus on pragmatic meanings contained in scientific texts (Norris
and Phillips 1994) was a switch to a focus on scientists' intentions. The upshot is
that scientific discourse requires interpretation that must take into account what the
writer or speaker intended, which is to say that the discourse is not totally independent of the context of its use. Thus, if the context is the description of methodology in a study, the scientific reader might be more inclined to take the text literally
than if the context is the interpretation of the results.

Over and above these psychological and epistemological connections between
literacy and science, scientists spend a great deal of their time engaged in literacy
activities. It could have been the case that, although necessary to science, scientists
spend but a small fraction of their time pursuing literacy tasks. The empirical data
show the opposite to be the case. Despite the data, "There is a widespread myth that
scholarly journals are seldom read" (Tenopir and King 2001, p. 114). Tenopir and
her colleagues trace this myth to a series of studies in the 1960s and 1970s funded
by the National Science Foundation in the United States. Many of these studies used
citation counts as evidence of frequency of reading, even though such counts were
known to be highly skewed towards a very small percentage of articles that are
heavily cited and away from the vast majority of articles that are cited infrequently
or not at all, even though they are read. Other studies failed to normalize data to
reflect reading done by the entire population from which a sample of scientists was
chosen. The error was shown to underestimate reading frequency by a factor as
large as 40 times.

In fact, "surveys of thousands of scientists from the 1970s, 1980s, and 1990s and
now into 2000 and 2001 consistently show that journal articles are considered to be
the most important information resource by scientists and are widely read" (Tenopir
and King 2001, p. 114). In a series of surveys beginning in 1977 and extending to
2005, Tenopir and King have shown that "university science faculty have increased
their number of readings in each survey time period observed ... University faculty
in 2005 report nearly twice as many readings as they did thirty years ago [280 v. 150
articles per year; 144 v. 120 hours per year]" (Tenopir and King 2008).

In addition to the fact that scientists read a great deal, the importance of reading
can be gauged by the amount of money that institutions employing scientists are
prepared to spend for this reading. The 144 h per year spent by each scientist
reading journal articles in 2005 amounts to a cost of thousands of dollars per year
for each scientist, measured in terms of their wages, an amount far in excess of the
cost of the journals themselves. We know also that scientists take their reading very
seriously: "In considering depth of reading, science faculty were most likely to read
a document 'with great care' or 'to get the main points' and far less likely to read a
document 'just to get the idea'" (Belefant-Miller and King 2000, p. 98).

Finally, the importance of reading is demonstrated by the fact that there is a
positive correlation between the number of journal articles read and the professional achievement of scientists: "award winners read 53 % more articles than non-
winners. Scientists who serve on high-level projects or special committees read
around 21 % more journal articles than those scientists who do not. Scientists who

are considered high achievers by their peers read 59 % more articles than their colleagues even when other variables are held constant. University scientists who have won awards for their teaching and those who have won awards for their research read roughly 26 % and 33 % more articles, respectively, than their cohorts" (Tenopir and King 2002, p. 260). Scientists for whom keeping up to date is more important were documented in one study to read 50 % more articles per month than those for whom keeping up to date is of lesser importance (Jamali and Nicholas 2010).

So, then, we have identified two ways in which reading is important to science and to scientists. First, for psychological and epistemological reasons, a basis in literacy, and hence in texts and in reading, is part of the constitutive architecture of science. Second, and coincident with this tight relationship between science and literacy, scientists just happen to read a great deal. There is one final way that reading is important to scientists that derives from these former two: it is that literacy helps to define the sort of communal social practices in which scientists engage. Scientists (a) record, present, and re-present data; (b) encode and preserve accepted science for other scientists; (c) peer review the ideas of other scientists anywhere in the world; (d) critically reexamine ideas once published; (e) connect ideas that were developed previously to current ideas; (f) communicate scientific ideas with those whom they have never met, even (metaphorically) with those who did not live contemporaneously; (g) encode variant positions on interpretations and methods; and (h) focus concerted attention on a fixed set of ideas for the purpose of interpretation, prediction, explanation, or evaluation (Norris and Phillips 2003). Without text, these social practices that make science possible could not be engaged efficiently or, in many cases, at all. These activities involve constructing, interpreting, selecting, and critiquing texts and are as much a part of what scientists do as are collecting, interpreting, and challenging data. These activities with text are as constitutive of the communal social practices of science as are observation, measurement, and calculation.

Students in today's schools, colleges, and universities might understandably miss these points about the importance of reading for science and scientists. It is the empirical basis of science that is at the forefront of science curricula. Students are told often that scientists do experiments, but rarely that scientists read and write text a great deal of the time. Students are told frequently that scientists wonder about the universe, but rarely that scientists ponder phrasing that will capture what they mean. Students learn that scientists explore the world, but few learn that scientists also search for written expressions that will carry the level of exactness they intend. Students are taught that scientists are not always certain, but they hardly ever are taught that this uncertainty leads scientists to choose words carefully to distinguish the degrees of certainty they wish to express in papers that they write. Students learn that scientists try to figure things out, but do not often learn that among these things are the meanings of what other scientists have written. Even though students early on pick up the idea that scientists make discoveries, after many years of school science they have not figured out that scientists also make choices about what to read, and make further choices about how closely and critically to read (Norris and Phillips 2008). In this pattern of learning and not learning, one essential feature of science is emphasized while another is almost entirely overlooked. Perhaps, then, it is not surprising that, when asked to describe what scientists do or to draw a scientist, students rarely say that scientists read or draw a scientist reading. Mead and Métraux (1957) included in their composite image of a scientist that "He writes neatly in black notebooks" (p. 387), but noted that a related part of the image, "He is always reading a book" (p. 387) was considered by students as part of the "negative side of the image of the scientist" (p. 387). Chambers (1983) reported that, in comparison to other indicators of the standard image of a scientist (e.g., lab coat, eyeglasses, facial hair, scientific instruments and laboratory equipment), "Much less frequently, we found symbols of science as recorded knowledge—the scientist ... writing in a notebook ... or reading some sort of report" (pp. 259–260).

The Structure of Science Texts

Three Meanings of 'Structure'

For the purposes of this volume, we assign three meanings to the term "text structure". When it is important to make a distinction, we make it clear which of the meanings we intend. On some occasions, we use the term to refer to all of the ways in which the structure of scientific texts can be understood, even ways that go beyond the three we consider. On those occasions, we use the term generally, without qualifications. The first meaning refers to the texts' overall *organizational structure*. Is the text divided into sections? If yes, what are the sections and what is their order of presentation? Do the sections have varying prominence, as suggested, for instance, by varying lengths? A second meaning of text structure denotes the *goal-directed structure* of the text, often signaled by the rhetorical moves authors make in various sections of the text. Thus, for example, authors establish the nature of the problem that their research addresses, summarize previous research on the topic, describe the general approach and methods to be used in the current research, report and interpret findings, and cast doubt upon alternative interpretations of their findings. A third meaning of text structure refers to the logic of the reasoning contained in the text. We term this logic the *argumentative structure* of the text. Does the text present data to confirm or disconfirm a hypothesis? Does the paper make a theoretical argument for a conclusion, perhaps drawing upon pre-existing data where relevant? Does the paper make predictions for future studies to test?

We show in the following section that some scholars have looked at scientific texts as forms of persuasive communication and have attempted to identify the various types of rhetorical moves presented by scientific authors. Other scholars have focused on the cogency of scientific texts and therefore have concentrated on the pertinence of the semantic connections between their various parts. It is reasonable to assume that writers structure their texts in particular ways to achieve certain purposes, so other scholars have examined the internal goal-directedness of scientific papers. In a vein similar to the emphasis on semantic connections, some researchers have concentrated on the logical sequencing within science texts. Still others have examined the speech acts present in scientific writings, which is an approach similar to the one marked by attention to goal-directedness. We group all of these approaches to analyzing scientific texts under the three meanings for text structure described in the previous paragraph.

The Importance of Understanding Structure

There has been much research over several decades showing that the structure of text and knowledge of the structure affect the quality of readers' interpretations. Cook and Mayer (1988, p. 448) remarked that even though "well-written scientific text has structure, some otherwise skilled readers may not be aware of the structure." This short remark contains several interesting thoughts and implications.

First, the idea that scientific text can be well-written is not one that is prominently taught as part of the school curriculum. A widespread myth is that science students do not need to be able to write well. Skilled writing is the preserve of language and literature majors, the myth continues, and it is in those subjects that writing typically is taught, as if learning to write in one discipline is fully generalizable to others.

Second, it is implied that being aware of the structure of the text is important to readers. That is, there is something other than access to the words and the information needed to interpret a text. There is something about the structure of the text that must be appreciated for the text to be grasped. The use of headings is one tool that scientists use to mark the structure of their texts—to relate their overall organization. Yet, it is not sufficient to notice the headings, "Method" or "Results". The reader must also see the implication that in the former section the author is presenting data not interpreting them or arguing for their significance, and that in the latter section the author is simply providing an account of what was done. In both sections, the author is presenting what he or she takes to be factual and setting aside judgment for later. If the reader misses this point, the text likely will be misunderstood.

Third, there is the presupposition that the attribution of skilled reading must be conditioned on the type of text being read. A reader skilled in reading literature, say, will not necessarily be skilled at reading science. As well, Cook and Mayer imply that the fact that transfer from one text type to another is not automatic is not simply a matter of the substantive content of the texts, but also of their different structures.

Swales (1981) was a pioneer in identifying the rhetorical "moves" scientists employ at the beginning of papers. He looked at articles in the physical, biological, medical, and social sciences and concluded that four moves were used: establishing the field, summarizing previous research, preparing for present research, and introducing present research. Building upon the earlier work by Swales, Swales and Najjar (1987) examined 66 articles from the *Physical Review* and 44 articles from the *Journal of Educational Psychology* that were published over a 40 year period beginning in 1943. They concluded that there was a growing trend over the years in the *Physical Review* but not in *the Journal of Educational Psychology* for introductory moves also to include an announcement of principal findings. Such moves establish the goal-directedness of the text. These structural components rarely are indicated explicitly. For example, authors usually do not write introductions like: "In the next few paragraphs I will be establishing within which field I work." Rather, such structures must be inferred by the reader. Unless the reader recognizes that at one time the author is reviewing previous research and at another time introducing the present research, then the reader is going to be at a total loss about the author's goals and, consequently, the meaning of the paper. This example demonstrates the accuracy of Cook and Mayer's point that a skilled reader of non-scientific text might miss the rhetorical moves at the beginning of scientific papers because such moves are not common or present at all in the type of text with which the reader is familiar. One of us has witnessed such a phenomenon (Brill et al. 2004) when observing the thinking process through which two high school students went while reading APL: "Yael and Liat also confused a past experiment that was brought up in the introduction of the article with the main experiment shown in the article, which had been based on this past experiment." (p. 505)

Hutchins (1977) focused on the semantic connections among statements in scientific papers. He provided a reasoning schema with variables for the specific statements that scientists might make in offering cogent and persuasive arguments. In Hutchins' view, the persuasiveness of a case of scientific reasoning arises from its cogency, and not, say, from its rhetorical flair. The schema is as follows:

(it is generally held; X is true)
but

    because (it has been found that A, B, C)
    (the problem is Y)

Therefore (hypothesis Z is proposed)
If (we apply tests D, E, F)

    so that (we have results G, H, J)

then (Z is true)
and

    if (this 'proof' is valid)
    then (V may be true)

      and (W may be false).   (p. 26)

Overlap can be seen between Swales' and Hutchins' schemata. For example, the following parts of Hutchins' schema correspond to Swales' moves of establishing the field and summarizing previous research:

(it is generally held; X is true)
but

    because (it has been found that A, B, C)

The following elements of Hutchins' schema correspond to Swales' moves of preparing for present research and introducing present research:

(the problem is Y)
Therefore (hypothesis Z is proposed)

Nevertheless, Hutchins' work primarily is about the logic or argumentative structure of the text as crucial to understanding its structure.

Hutchins' work, although valuable for its insight, has been criticized (e.g., Crookes 1986) for its over-reliance on Gopnik (1972) who studied abstracts of papers rather than complete scientific papers. Furthermore, the schema represents thinking associated with proving or disproving hypotheses. Although hypothesis testing frequently takes place and is important in science, not all of scientific work involves the testing of hypotheses. Nevertheless, the above schema is instructive because it shows a textual schema that would make concrete sense if the variables were instantiated. A student who grasped the import of the schema would have knowledge of what Norris (1992, p. 216) called "the general shape that a justification would have to take, of the general sorts of considerations that scientists would count as justification, not knowledge of the justification's actual details." Thus, there is an aspect to the structure of scientific texts the understanding of which demonstrates knowledge of the nature of science. Even if the schema proposed by Hutchins does not cover all types of scientific texts, and even if it does not completely accurately capture the form it is about, namely, reasoning about hypotheses, it points to an overarching truth. That truth is that, although students cannot grasp many of the particulars of scientific text, there is still something general for them to learn about the argumentative structure of scientific text that provides insight into the nature of science.

Suppe (1998), a philosopher of science, examined more than 1,000 data-based scientific papers. When analyzing structure, he focused on what the scientists were attempting to achieve in their papers, that is, their goal-directedness. He identified several goals in the scientists' writing: presenting data from an observation or series of observations; making a case for why the observations were relevant to a particular scientific problem; providing details on the methods used to collect and analyze the data; providing and justifying interpretations of the data; and addressing specific doubts arising from alternative interpretations of the data. These scientific goals are interesting from the point of view of school science instruction because several of them receive hardly any or no attention in the science curriculum. For instance, the goals of justifying the relevance of observations and the interpretations of data, and dealing with alternative interpretations are far distant from most school science curricula.

Section Summary

In this section, we have identified three types of structure found in scientific texts: organizational, goal-directed, and argumentative. We have reviewed some of the scholarship dealing with each of these categories, which exemplifies different approaches to dealing with each type of structure. We have not attempted to be exhaustive in examining all possible ways to consider the structure of scientific text. Our main goal was to highlight some very important features of structure and to make an important point: namely, if a reader wishes to interpret and to any extent grasp the meaning of a scientific text, then these aspects of the text's structure first must be recognized and understood.

Contrasting Examples: The Recognition of Structure and the Determination of Meaning

Contrasts Between Experimental and Theoretical Papers

For the purpose of providing concrete examples, we conducted our own examination of data-based research papers by colleagues in physics (Norris and Phillips 2008). One paper reported a study of hysteresis. In hysteresis, a system undergoes a change of state in response to changes in its environment that does not reverse when the environment reverts to its original state. For example, glass thermometers, once heated and having undergone expansion, do not return to their original sizes when their environments return to their original temperatures. If we magnetize a piece of iron by placing it in the magnetic field, removing the magnetic field does not lead to the piece of iron demagnetizing by simply reversing the route of its magnetization. In order to demagnetize the iron, energy has to be expended in the form of a magnetic field of opposite polarity to the one that originally induced the magnetism in the iron. The states of some physical systems are thus dependent not only upon the current states of their environments but also upon their histories—the environments in which they have been. So it is, apparently, with the adsorption and desorption of helium in silica aerogels, a type of porous media studied in condensed matter physics (Beamish and Herman 2003).

Beamish's and Herman's paper of just two pages exhibits many of the goal-directed elements identified by Suppe. Beamish and Herman:

1. motivated their study—"Silica aerogels ... provide a unique opportunity to study the effects of disorder on phase transitions" (p. 340).
2. reported relevant past results—"Also, recent experiments ... showed features characteristic of capillary condensation rather than true two-phase coexistence ..." (p. 340).

3. reported limitations of past research—"... although long thermal time constants made it difficult to determine the equilibrium behavior" (p. 340).
4. described what was done—"... a thin (0.6 mm) disc was cut ... Copper electrodes (9 mm diameter) were evaporated ... This was sealed into a copper cell ... Temperatures were measured and controlled ... A room temperature gas handling system and flow controller allowed us to admit or remove helium at controlled rates..." (p. 340).
5. argued for the suitability of techniques—"This effect [thermal lags when gas was admitted or removed] appears as a rate dependent hysteresis, so it was essential to check directly for equilibrium" (p. 341).
6. explained observations—"This behavior [The equilibrium isotherm ... had a small but reproducible hysteresis loop and did not exhibit the sharp vertical step characteristic of two phase coexistence] is characteristic of surface tension driven capillary condensation in a porous medium with a narrow range of pore sizes" (p. 341).
7. conjectured what might be happening—"At low temperatures we observed hysteresis between filling and emptying, as expected for capillary condensation; this hysteresis disappeared above 5.155 K. The vanishing of the hysteresis may correspond to the critical temperature of the confined fluid" (p. 341).
8. challenged interpretations—"However, this identification [with the critical temperature] is not clear, since all our isotherms have finite slopes [while theory would suggest an infinite slope at this point]" (p. 341).

The second paper reported a study of the transition from crystalline to amorphous solids in certain nanocrystals under the influence of radiation (Meldrum et al. 2002). In comparison to crystalline solids, amorphous solids (a common example is glass) do not have their molecules arranged in an orderly geometric pattern. They have no definite melting or freezing points (when heated, they turn soft before becoming liquid; unlike ice, for instance, which jumps from solid to liquid without passing through an intermediary phase), usually have high viscosities (glass runs very slowly), and fracture along irregular patterns rather than regular cleavage lines. We exemplify several sections of the paper that reveal its argumentative structure.

1. As part of the motivation of the study, whose aim was to provide evidence for the amorphization of nanocrystalline zirconia by ion irradiation, the authors provided a full-paragraph case for the proposition that zirconia is one of the most radiation resistant ceramics known. The point of the argument was to signal that the results of this study were not to be expected, that previous conclusions will have to be modified, and that therefore this study was significant.
2. In addition to describing what was done to collect data, two justifications were provided for the methods chosen, which were deemed particularly appropriate in that they avoided certain pitfalls that would have led to misleading results.
3. Imbedded in the description of the results, a case was made that the crystals of zirconia that reached the amorphous state were sufficiently small to allow the formation of tetragonal zirconia. However, in the case under study, there was

incomplete information for calculating the critical size for such formation. By comparing two related cases where the relevant calculations could be made, the authors argued that the size of their crystals was smaller than either of these two cases and that therefore they should be sufficiently small to support the observations made.

4. These authors spent considerable effort arguing against alternative interpretations of the observations they made. They carefully spelled out how various steps of their method made other possible mechanisms implausible.

Of course, not all scientific papers report the results of empirical studies. Harmon (1992) studied the contents of theoretical science papers. He looked at the 400 most cited theoretical papers between the years 1945–1988. Compared to experimental papers, these papers contained fewer tables and figures, but a far great number of equations and references. The papers were described as following this sequence:

> ... problem or need, assumptions made in attempting to solve problem or meet need, theorem derived from those assumptions and additional considerations, proof of theorem by logical reasoning or validation by comparison with what is established or establishable, conclusions from previous discussion, and recommendations on future experimental or theoretical work. (p. 357).

From this brief account, we see that, similar to data-based papers, theoretical papers have an organizational structure, a goal-directedness, and argumentation. One feature of theoretical papers that sets them apart from experimental and other data-driven papers is that in the latter "we find facts and events that normally take place in the real world" (Horsella and Sindermann 1992, p. 131) compared to the former in which we find descriptions of "situations ... that take place hypothetically in [sometimes] counterfactual worlds" (Horsella and Sindermann 1992, p. 131). We have added the word "sometimes" because we do not believe that all theoretical thinking is counterfactual in imagining what would be the case if some condition known or thought to be true were not true. Sometimes theoretical thinking proceeds in this way, but on other occasions it straightforwardly deals with questions of what is the case to make some known condition occur.

We have provided several examples of how a recognition of structure affords access to the meaning of scientific texts and to the nature of science, even in the absence of knowledge of many of the technical scientific details. We return to these examples in Chap. 3 to make the case that APL is a powerful resource for teaching students to recognize the structure of scientific text.

Contrasts Between and Within Disciplines

We examine a within-discipline difference first by contrasting Beamish and Herman's paper on silica aerogels to a paper on the anomalous hotness of the Sun's corona from *The Astrophysical Journal*. For the sake of this analysis, we will say that both of these papers fall within the broad discipline of Physics, recognizing the disciplinary boundaries are impossible to draw sharply. Second, we describe three contrasts between disciplines: an experimental biology paper to the two physics papers, the theoretical physics paper to a paper in mathematical biology, and between the theoretical physics paper and a similarly structured paper in archaeology. The section is about how the structural elements (organization, goal-directedness, argumentative) play out in actual scientific papers. The point is to show that there is no rigid structural patterning—there is as much variation both within and between disciplines.

Within-Discipline Contrast

The Beamish and Herman paper is, as we have said, two pages in length. Its organizational structure is defined by five sections: Abstract, Introduction and Experimental Details, Results and Discussion, Acknowledgement, and References. The first two sections take up most of the first page; the results and discussion occupy most of the second page, with about two-thirds of the space dedicated to reporting results. The acknowledgements' section is one line long, and there are five references. The paper is a concentrated version of a longer experimental study report. Data are reported graphically only, which seems sufficient to demonstrate the hysteresis loops. A reader wishing to have the exact data points would have to contact the corresponding author. The main purpose of the paper is to report what was observed. The goal-directed structure is displayed by the relatively large amount of space devoted to results. Only one sentence is devoted to offering a tentative explanation of one of the observed results (the vanishing of hysteresis above $5.155°K$). The paper wastes no space and deals only with the bare essentials. It contains 1,193 words.

In contrast, the paper by Aschwanden et al. (2007) that appeared in *The Astro-physical Journal* is 9 pages long and contains 6,773 words. The headings are: Abstract, Introduction, Coronal Heating Requirement, Ten arguments for Heating in the Chromosphere/Transition Region, (10 subheadings, one for each of the 10 arguments), Conclusions, Appendix, References. There are 67 references listed in this paper. The section containing the 10 arguments is by far the longest in the paper, comprising approximately 5 of the 9 pages, which is consistent with the goal of the paper to make a case for why the Sun's corona is so hot. The paper contains 7 figures. Two of the figures depict contrasting models of coronal and chromo-spheric heating. The remaining 5 figures present data. These are not data collected in the current study, but data that had been reported in previous papers that was judged pertinent to the case being made. The argumentative structure of the paper is explicit, as is indicated by the heading of the third section. Thus, although falling into the broad discipline of physics like the Beamish and Herman article, the paper has a different organizational structure, goal-directedness, and argumentative struc-ture. Rather than an exercise in reporting new observations, the paper relies upon already known observations to construct an argument for a novel conclusion. The data were not necessarily gathered in order to support this argument. Indeed, it seems that none of the data were amassed for this specific purpose. Instead, the authors opportunistically gathered data from a variety of other studies to make a case for which perhaps none of the authors of those other studies even anticipated their data being used.

Between-Discipline Contrasts

First, we examine the report of an experimental study, which, because of its experimental nature, resembles the report by Beamish and Herman, but it is in the field of biology. The paper appeared in *Nature Biotechnology* (Mourez et al. 2001). It runs just under 4 pages in length and contains 4,149 words. The paper contains 1 table and 3 figures, one of which depicts a chemical mechanism and two of which report results. The sections in the paper are as follows: An abstract that is not labeled as such; an introduction that is not labeled as such; Results; Discussion; Experimental Protocol; and references, which are not labeled as such. Although an experimental study like the one by Beamish and Herman, this paper has a different organizational structure. The results are presented right after the introduction and before a detailed description of the methods in the experimental protocol section. Having said that, the results section does contain brief descriptions of the methods alongside the results, and also brief statements of theory and references to results of other studies. The methods are described most fully right at the end of the paper, before the references. Unlike the Beamish and Herman and Aschwanden et al. papers, the discussion section contains some commentary on applications, which were foreshadowed in the second sentence of the paper, which referred to "biological warfare and terrorism" (p. 958).

The second contrast is between the Aschwanden et al. paper and a paper in mathematical biology that appeared in the *Proceedings of the Royal Society B* (Wonham et al. 2004). The paper is dissimilar to Aschwanden et al.'s in that the primary disciplines are mathematics and biology, compared to physics and astrophysics. The paper is similar in that it does not present new data but draws on data published in other papers or made available by scientific agencies. The paper also advances a theoretical model and defends it; in this case the model describes the transmission of the West Nile virus between birds and mosquitoes. The papers are also similar in that they both muster arguments in defense of their models, although, considering organizational structure, the arguments are put forward more explicitly in the case of Aschwanden et al. The Wonham et al. paper is 7 pages long, comprising 5,695 words. The paper contains the following sections: An abstract, which is not labeled as a separate section; Introduction; Model Description; Model Analysis; Model Predictions; Public Health Implications; Temporal Extensions; and References. The reference list is 35 items long. The description of the model and its analysis occupies about 3 pages, or about 43 % of the paper compared to about 55 % for the longest section in the Aschwanden et al. paper. What is strikingly different about the Wonham et al. paper compared to the Aschwanden et al. paper is the amount of space (just over 2 of the 7 pages or about 28 % of the paper) devoted to implications for practice, in this case dealing with the diminution of mosquito populations. This contrast in goal-directedness might be explained by the fact that the topic of the Wonham et al. paper leads fairly naturally to practical application compared to both of the other two papers, which are removed substantially from practical application. As an aside, although the Mourez et al. paper included some references to practical applications of their research, they occupied a much smaller proportion of space than was devoted to applications in the Wonham et al. paper.

The third contrast is between an archaeology paper and the Aschwanden et al. paper in astrophysics. Bell and Renouf published a paper in *World Archaeology* in 2004 arguing

> that variable and complex post-glacial relative sea level (RSL) in Newfoundland is linked to (1) the uneven distribution around the coastline of late Maritime Archaic Indian (MAI) sites (5500–3200 BP) and (2) the apparent absence of early MAI sites (8000–5500 BP), despite their presence in nearby southern Labrador. (p. 350)

Both the archaeology paper and the astrophysics paper present arguments that rely on data from previous studies. The Bell and Renouf paper is 21 pages in length and contains 8,117 words. The paper has the following sections: Abstract; Keywords; Introduction; Archaeological Context; Relative Sea Level History; Late MAI Sites and RSL History; Early MAI Sites and RSL History, Early MAI Sites on the Northern Peninsula; Early MAI Site Preferences; Summary and Conclusion; Acknowledgements; Notes; and References, containing 47 items. The paper contains 2 tables and 9 figures. Each figure includes an explanatory caption, and in several cases the captions are extensive. The paper attempts to explain two facts: (1) that late MAI sites are distributed unevenly around the current coastline of Newfoundland; and (2) that early MAI sites are completely absent, even though many such sites have been found on the coast of Labrador, from which Newfoundland is visible along much of the Strait of Belle Isle. The authors used existing published physical geography data on where the coastline was on either side of three glacial periods, and demonstrated a complex pattern of coastal emergence and submergence. Bell and Renouf argued that, with regards to the current coastline where late MAI sites are not found, these sites might be found offshore. Indeed, one such site is known to be exposed only at low tide and two others at 1 and 5 m below lowest sea level. They predicted that other sites would be found submerged under 20 or more meters of ocean water. With regards to early MAI sites they argued that, depending upon the section of today's coastline being studied, some sites should be found in the shallows offshore and other sites should be discovered onshore well above today's sea level. This latter explanation for the absence of early MAI sites has yet to be confirmed or disconfirmed by the discovery or failure to discover such sites well inland of today's coastline.

Like the Aschwanden et al. paper, the Bell and Renouf paper contains no methods section. Unlike the Aschwanden et al. paper, which contains a specific section with 10 labeled arguments for the conclusion that heating takes place in the chromosphere, the Bell and Renouf paper distributes their arguments throughout the paper and does not refer to them explicitly with heading labels. Although they are from distinctly different disciplines, the papers are alike in that they begin from an accepted premise that something is the case: in the Aschwanden et al. case it is accepted without question for the purposes of the paper that the Sun's corona is anomalously hot; in the Bell and Renouf paper it is accepted without question for the purposes of the paper that the late MAI sites are distributed anomalously unevenly around the current coastline of Newfoundland and that early MAI sites are unusually absent altogether. In each paper, the authors muster other accepted facts to construct and propose an explanation of the anomalies.

Section Discussion and Summary

These contrasting examples suggest that the organizational, goal-directed, and argumentative structures of scientific papers can vary considerably both within and between fields. Often, the structure is related to purpose: for example, the Aschwanden et al., Bell and Renouf, and Wonham et al. papers concentrate on making cases to account for known facts. This goal drives the organizational structures of the papers far more than disciplinary affiliation. In view of the fact that this was the goal in each case, it is directly on that goal that most effort was expended. Yet, papers with very similar goals, such as the purpose of the Beamish and Herman and the Mourez et al. papers to report new empirical findings, can have quite different organizational structures. The lesson would seem to be that one's beliefs about the structure of scientific text ought not to be too rigid: scientific texts vary in structure for a myriad of reasons including the goals of the research being reported, the scientific discipline, and perhaps even the style of communication preferred by the scientists or the publishing journal.

Some might question whether papers that "concentrate on making cases to account for known facts" should be considered primary scientific literature at all. A case can be made to restrict use of the term "primary" to refer to reporting original discoveries for the first time. When others subsequently use these discoveries, as often occurs in theoretical papers, then the papers no longer are primary. At least this is a case that might be made. Yet, the authors in the Coronal Heating Paradox paper, for example, bring new arguments to known facts and this can be a reason to consider their paper as primary literature. According to the society that published the article

> Papers published in *The Astrophysical Journal* present the results of significant original research not previously published. Articles submitted to the Journal should meet this criterion and must not be under consideration for publication elsewhere. Commentary on previously published papers does not constitute significant original research. (American Astronomical Society 2013)

That is, the Society distinguishes between commentary on previously published work, which is not considered primary, and work that draws on the data of previously published work to "present the results of significant original research not previously published".

Similarly, for the West Nile virus paper, the authors take known facts in order to build their mathematical model, which is their contribution to the field. According to the Royal Society that published their paper, "Proceedings B is the Royal Society's flagship biological research journal" (Royal Society 2013) for which

> The criteria for selection are: work of outstanding importance, scientific excellence, originality and potential interest to a wide spectrum of biologists. ... Submission of preliminary reports, of articles that merely confirm previous findings, and of articles that are likely to interest only small groups of specialists, is not encouraged. (Royal Society 2013)

That is, the Royal Society in publishing the paper judged it as important, excellent, and, most significantly, original. These characteristics qualify it as an instance of primary scientific literature.

We note that the structural differences that we identified arise for a variety of reasons. Even though structure is a guide to interpretation, it is likely that for experts in a field variations in organizational structure do not present problems for interpretation. However, for novices such variations might be confusing and there might be some organizations that are easier for novices to understand than other organizations. Since APL articles are not written for experts, we added headings (Introduction, Methods, etc.) when they were not present in the originals in order to make the structure more apparent to the novice readers. Also note that some sections, such as the methods section, appeared in various locations in different journals, sometimes without a label. We do not see any problem with these variations for experts. We believe it is important for novices to understand the methodology before reading the results, so we both label the sections and place the methods first.

Representation of Epistemology in Scientific Texts

The scientific papers that we have been describing make possible a group of communal social practices that combine to produce science. We have listed some of these practices in the first section of this chapter as recording data, peer reviewing, communicating ideas, and so on. Underlying these practices is a key ingredient: all the papers have an argumentative structure in the sense of providing reasons, data, and evidence for conclusions. Another similarity we wish to examine is that all of the papers express a degree of tentativeness in their findings. These characteristics define the epistemology upon which the papers are based. We call this epistemology, "fallible rationality". The epistemology is rational in that it demands a reliance upon reasons: conclusions are expected to conform to the reasons available for supporting them. We use the word "conform"—meaning to bring into harmony or accord—because it expresses the level of precision suitable to the relationship. Reasons do not always compel scientific conclusions, though sometimes the evidence is so strong that it makes a specific conclusion inevitable. At other times, the evidence is strong, but not irrefutably so. At yet other times, the evidence is indicative but cannot be considered strong. Thus, the scope and certainty of a conclusion must be tailored to match the breadth and the strength of the reasons and evidence that can be mustered on its behalf. According to this epistemology the strength of conclusions is directly linked to the strength of the reasons and evidence available to support them. There hardly ever is an algorithm for computing these strengths—expert judgment is involved, which in science typically involves a community of experts in a given area.

The epistemology is fallible because no matter how conclusive the currently available evidence, there is always the possibility of finding additional refuting evidence. According to this epistemology, evidence can cut both ways—it can both corroborate conclusions and disconfirm them—and there can be no prior knowledge of what evidence will come to light. According to this epistemology, there is no such thing as complete evidence, that is, a point at which all relevant evidence has been gathered and no additional relevant evidence is ever possible. According to this epistemology, not only is additional evidence relevant by current conceptions always conceivable, the very idea of what evidence is relevant to a conclusion can change with time. Thus, evidence can be admitted as relevant that once was thought not to be so, and new evidence not even imagined under previous conceptions of relevance can be brought into consideration.

Reliance on a fallible rationality is evident in each of the five papers we have analyzed. Consider the two experimental papers first. Mourez et al. states the following at the opening of the discussion section about the success of their experiment in blocking the toxicity of anthrax:

> The blockage of toxicity observed in our studies occurred by specific interaction of PVI with PA63 generated *in vivo* on host cells, and is not due to an interaction of PVI with native PA or LF in solution. This conclusion is supported by the finding that PVI affected the electrophoretic mobility only of PA63 . . . . (p. 960).

Even without understanding the technicalities, it is clear from the language that the authors were defending their conclusion on the basis of data offered in evidence, which satisfies the rationalist characteristic of the epistemology we claim underlies each of the papers. Note the use of the word "supported" for referring to the relation between the evidence and the conclusion, as distinct, say, from the word "proven." This distinction shows the scientists to be fallibilists. "Support" is used in the sense of "uphold" and "defend", which leaves open the possibility of other findings that do the opposite.

The Beamish and Herman paper exhibits the rationalist component of the epistemology by constantly referring to their data in drawing their conclusion that at "low temperatures we observed hysteresis between filling and emptying" (p. 341). They demonstrate the fallibilist side of their epistemology in their interpretation of the finding that the "hysteresis disappeared about 5.155 K" (p. 341). They offer the idea that the "vanishing of the hysteresis may correspond to the critical temperature of the confined liquid" but that "this identification is not clear, since all our isotherms have finite slopes, while ... at a liquid-vapor critical point would correspond to infinite slope" (p. 341). In this passage, the authors both conjecture an explanation of one of their results and provide the argument to undermine it, suggesting clearly the implication that the interpretation is tentative and liable to revision.

Now, let us consider the two papers that rely upon data obtained by previous studies. Bell and Renouf are the most explicit in indicating their epistemology. They use the following words or their cognates over 25 times: "argue", "because", "conclude", "probably" (in the sense of "insofar as seems reasonably true"), "roughly" (in the sense of "approximately"), and "suggest". Their word choice undergirds their fallible rational epistemology, as demonstrated in the following passage:

> Understanding the changing post-glacial coastline is crucial in regions where prehistoric human occupation was tied to the coast. Newfoundland is such a region. In this paper we have argued for a link between the uneven distribution of late MAI sites (5500–3200 BP) throughout the coastal regions of Newfoundland and Newfoundland's variable and complex RSL history. We argue that, in those regions where there has been considerable post-glacial submergence, sites earlier than 3000 BP will be submerged in the shallow offshore. We predict that sites older than 3000 BP will be submerged in areas where the lowstand is greater than 20 m. This would explain the absence of known MAI sites on the south-west coast and southern Avalon Peninsula and the relative absence of sites along the south coast. (p. 366)

In this passage, the authors show their reliance on evidence for drawing tentative conclusions. The evidence pertains to two phenomena: the variable relative sea level over the past 10,000 years in Newfoundland, and the uneven distribution of late Maritime Archaic Indian sites around the current coast of the island. The first phenomenon is used to offer an explanation of the second one, by proposing that many sites currently either are submerged or sit high above current sea level. Twice in this short passage, they make clear that they are arguing for a link between the two phenomena, making it explicit that they are leaving it open to the reader to accept their argument, reject it, or to hold it in abeyance awaiting additional evidence. In the final sentence, they also write using the auxiliary verb "would" to express a possibility rather than a certainty. That is, *if* indeed the late MAI sites were positioned near the coast between 5,500 and 3,200 years ago, then the variable sea level would explain why they are not found today in greater and more evenly distributed abundance.

Moving to the Aschwanden et al. paper, we see the authors' epistemology revealed in the title to section 3: "Ten arguments for heating in the chromosphere/transition region" (p. 659). It is clear from beginning to end that the authors are attempting to marshal sufficient evidence to construct a convincing case for their contention that "coronal heating" is a misnomer for a phenomenon that really occurs in the sun's upper chromosphere. The following passage taken from the conclusions section demonstrates the fallible rationality character of the authors' epistemology:

We have enumerated 10 arguments that the primary heating process leading to the appearance of hot (T ≥ 1.0 MK) coronal loops takes place in the chromosphere rather than in the corona. Most of the arguments are based on observational constraints that are often ignored in theoretical models: [10 constraints are specified one after another] ... Given these 10 arguments, we see strong support for the working hypothesis that the major heating source responsible for a hot corona is located in the transition region rather than in the corona. We therefore consider the term "coronal heating" as paradoxical, since the essential heating process does not take place in the corona, but rather in the transition region or upper chromosphere. ... This paradigm shift of the heating source from the corona to the chromosphere does not, however, exclude that the primary energy release could occur in the corona ... (pp. 1679–1680)

First, the authors make explicit use of the word "argument" and say that in these arguments they "see strong support for the working hypothesis" that the heating takes place outside the corona. They make clear the source of the reasons in their arguments—the ten observational constraints that often are ignored by others. It is indicative of their acceptance of fallibility that they use the expression, "we see strong support", which implicitly acknowledges that not everyone would evaluate their arguments as positively as they do. They also try in the final sentence to set fairly the bounds of their conclusion by stating what the conclusion does not exclude, namely, the primary energy release occurring in the corona. In specifying the boundaries so clearly, they once more recognize the fallibility of their conclusions.

We turn finally to the Wonham et al. article on West Nile virus transmission. Much of the article is devoted to developing a mathematical model. The reasoning frequently is deductive, so that as long as the premises are true and the deductions valid the conclusions hold without doubt. The reliance on a deductive link to conclusions is different from the evidence-to-conclusion link found in the other four papers. The deductions make the reasoning appear less susceptible to challenge and the conclusions more secure. At the same time, the authors signal their reasoning, with seven uses in total of the words "therefore" and "since".

The authors clearly are fallibilists. Consider first of all the number of times they use "assume" or a cognate, which is five times. Each time they announce an assumption, they are inviting their readers to go along with their thinking. Of course, such invitations can be turned down, in which case the deductions based upon the assumptions do not go through. Depending upon the centrality of the assumptions to the model, then the ramifications for their overall case vary in the intensity of negative effect. They also hedge their argument with terms such as "generally considered", "we believe", "would probably be impossible" remarked about total West Nile virus eradication which is allowed under their model, and "an important first step" stated in reference to their model.

In this section, we have made the case that the scientific papers we have examined are based upon an epistemology characterized by rationality—the giving of reasons in the defense of conclusions—and fallibility—the liability to error and the tentativeness that its recognition evokes. For each paper, we have called attention to the authors' reliance on data and on theoretical and logical argument to support their conclusions. We have brought into view the authors' widespread use of specific language qualifiers to signal that their conclusions were offered tentatively.

Meta-scientific Language

Meta-scientific language is designed to deal critically with scientific language and, derivatively, scientific practice. Unlike the terms discussed in the previous section, which are markers for rationality and fallibility, meta-scientific language enables speakers and writers to refer to scientific practices and to talk about them in a theoretical and critical way. It is widely recognized that historians, philosophers, and sociologists of science talk and write about science by employing a meta-scientific language that affords them a stance as an observer of science. That is, their stance is as examiners of science from the outside-in. In their papers, scientists also employ a meta-scientific language that enables them, from time-to-time, to adopt an outsider's stance even while they remain insiders to their research.

First, let us note that scientific language is textured in that not all of its statements have the same reported or implied truth status (Norris and Phillips 1994). We see this texture in the papers examined. There are statements offered as truths, for example, descriptions of what has been found in the past, of what was done in the current study, and of what was observed. There are statements offered as probable or as having uncertain truth status, for example, conjectured explanations. There are statements offered as false, for example, explanations that the authors aim to impeach. In all of these cases, the statements are offered from the point of view of the insider conducting the research and presenting its findings.

Scientific language also is structured by use of a meta-scientific language that includes such terms as "cause", "effect", "observe", "hypothesis", "data", "results", "explanation", and "prediction". Table 3.1 indicates which of the meta-scientific terms from our list are found in the five articles we have examined.

Each of the articles, with the exception of Beamish and Herman, used at least six of the eight terms. Most terms, when used, were used multiple times. Perhaps most significant is the fact that three terms were used in all five papers: "observe", "data", and "results". All three of these terms refer to evidence in one way or another, emphasizing the rational character of the authors' epistemologies.

Each of the meta-scientific language terms implies a relationship and an evaluation based upon that relationship. We take one of them, "observe", and one of the articles, Aschwanden et al., to show what we mean. The first use of "observe" or a cognate is in the abstract: "In this paper we point out that observations show no evidence for local heating in the solar corona" (Aschwanden et al., p. 1673). The statement presupposes its readers understand two antecedent points. The first point is that observations can stand in the relationship of evidence to a claim or hypothesis. In this case, the hypothesis is that the solar corona derives its heat from a mechanism within itself. The second point is that referring to a phenomenon as observed is to evaluate it as suitable for providing evidence.

Let us examine how Aschwanden and his colleagues use these antecedent points at several locations throughout their article. In the first argument comparing the coronal heating to the chromospheric heating models, the authors claim in the caption to a figure depicting both models that "The heating phase of the coronal

Table 3.1 Meta-scientific language occurrence in five scientific articles

Meta-language terms and cognates	Citation to scientific article				
	Mourez et al. (2001)	Beamish and Herman (2003)	Bell and Renouf (2004)	Aschwanden et al. (2007)	Wonham et al. (2004)
Cause	✔		✔	✔	✔
Effect	✔	✔		✔	✔
Observe	✔	✔	✔	✔	✔
Hypothesis	✔			✔	
Data	✔	✔	✔	✔	✔
Results	✔	✔	✔	✔	✔
Explain			✔	✔	
Predict			✔	✔	✔

✔ = Term or cognate occurs

heating scenario should be observable" (p. 1674). In the text they claim that such a "heating phase . . . has never been observed" (p. 1674). In this reasoning, observation is being used evaluatively, in the sense that observations are taken to be suitable for supporting ideas and in the sense that lack of relevant observations is taken as an impeachment of an idea. "Observation" is not simply a neutral term for describing a scientific practice but a value term—to say that something is an observation is to assign it value as something that is reliable and as something upon which scientists should rely (Norris 1985). This set of ideas resides outside any scientific theory and is not cast in the language of any scientific theory. It is cast in meta-scientific language that enables Aschwanden et al. to reflect upon and evaluate scientific practice.

Here are several more examples from the Aschwanden et al. article that use observation in the same meta-scientific sense:

> The best direct evidence for chromospheric heating with subsequent coronal filling is observations of hot upflows. (p. 1675)
> Therefore, steady upflows can keep radiatively cooling loops at nearly constant temperatures (in time) as observed. (p. 1677)
> In the model of Parker . . . heated plasma would spread along the coronal field lines and should manifest itself in long (probably unresolved) thin threads, for which we have no observational evidence. (p. 1677)
> This observational result is hard to reconcile with any nanoflare model that predicts unresolved loop strands. (p. 1678)
> We have enumerated 10 arguments. . . . Most of the arguments are based on observational constraints that are often ignored in theoretical models. (p. 1679)

It is clear in each case that pointing out that something is an observation or that some event has never been observed is not simply to make a descriptive claim. The point in each case is to make an evaluation, either that an idea for which the observation is taken to be relevant evidence either should be accepted or rejected.

Concluding Comments

Myers, who has done an enormous amount of research into the nature of scientific text, makes a very interesting point that relates to our ideas of texture and structure. Non-specialists faced with scientific texts often interpret their difficulty in understanding as one of not knowing the vocabulary. However, argues Myers, the problem will not go away by using a dictionary, no matter how good or specialized, because much of the difficulty interpreting scientific text lies in grasping the connections of one statement to another (Myers 1991). As we have just finished pointing out, one way in which such connections are traced is through the use of meta-scientific language. That is, it is important for students of science to understand the meanings scientists intend to convey by keys terms such as "cause", "observe", "hypothesis", and "explain". Yet, there is much more involved in grasping how scientific texts are knitted together.

First, recognizing that science has a literacy as well as an empirical basis is key. The recognition that science has a literacy basis forces the realization that scientific text requires interpretation, contrary to the belief that it wears meaning on its face. This realization is precisely the same as the one that advocates of an empirically grounded view of science have urged for decades: namely, that the world does not wear meaning on its face but must be interpreted through the intellectual tools developed by science. Therefore, the literacy basis of science and the empirical basis of science are similarly motivated and entirely compatible viewpoints about science.

Second, the connections that bind scientific texts together can be revealed through an understanding of the structure of the text, its organizational, goal-directed, and argumentative structure. The central idea is that the science student requires access to more than the meaning of the words in order to lay hold with the mind to relationships implicit in how the parts of the text are organized.

Third, students must comprehend and learn to accept the tension inherent in the epistemology of science. At one and the same time, scientists struggle to put forward unassailable evidence for claims all the while acknowledging that they are capable, and even liable, to make mistakes given their best efforts. Societies as a whole have not fully come to grips with these countervailing tendencies within science, which might mark one of the major failings of science education. The fallible rationality of science is totally contrary to black and white thinking. Thus, although it is natural for scientists to admit mistakes and to change their minds, such behavior often is seen by those outside of science as weakness. Within science, such behavior is admired and is recognized as the only foundation that science has, as shaky as it might be. Science texts, and well-designed APL, contain this tension by the very nature of their genre.

Acknowledgements Parts of this chapter draw upon the following previously published work:

Norris, S. P., & Phillips, L. M. (2008). Reading as inquiry. In R. A. Duschl & R. E. Grandy (Eds.), *Teaching scientific inquiry: Recommendations for research and implementation* (pp. 233–262). Rotterdam: Sense.

Norris, S. P., & Phillips. L. M. (2003). How literacy in its fundamental sense is central to scientific literacy. *Science Education, 87*, 224–240.

We thank the publishers of these works for permission to re-publish.

References

Aschwanden, M. J., Winebarger, A., Tsiklauri, D., & Peter, H. (2007). The coronal heating paradox. *Astrophysical Journal, 659*, 1673–1681.

American Astronomical Society. (2013). *The Astrophysical Journal.* Retrieved June 4, 2013, from http://iopscience.iop.org/0004-637X/page/Scope

Beamish, J., & Herman, T. (2003). Adsorption and desorption of helium in aerogels. *Physica B, 329–333*, 340–341.

Belefant-Miller, H., & King, D. W. (2000). How, what, and why science faculty read. *Science and Technology Libraries, 19*, 91–112.

Bell, T., & Renouf, M. A. P. (2004). Prehistoric cultures, reconstructed coasts: Maritime Archaic Indian site distribution in Newfoundland. *World Archaeology, 35*(3), 350–370.

Brill, G., Falk, H., & Yarden, A. (2004). The learning processes of two high-school biology students when reading primary literature. *International Journal of Science Education, 26*, 497–512.

Chambers, D. W. (1983). Stereotypic images of the scientist: The draw a scientist test. *Science Education, 67*, 255–265.

Cook, L. K., & Mayer, R. E. (1988). Teaching readers about the structure of scientific text. *Journal of Educational Psychology, 80*, 448–456.

Crookes, G. (1986). Towards a validated analysis of scientific text structure. *Applied Linguistics, 7*, 57–70.

Finson, K. D. (2003). Applicability of the DAST-C to the images of scientists drawn by students of different racial groups. *Journal of Elementary Science Education, 15*(1), 15–26.

Fung, Y. Y. H. (2002). A comparative study of primary and secondary school students' images of scientists. *Research in Science and Technological Education, 20*, 199–213.

Gopnik, M. (1972). *Linguistic structures in scientific texts* (Janua Linguarum, Series minor, 129). The Hague: Mouton.

Harmon, J. E. (1992). Current contents of theoretical scientific papers. *Journal of Technical Writing and Communication, 22*, 357–375.

Horsella, M., & Sindermann, G. (1992). Aspects of scientific discourse: Conditional argumentation. *English for Specific Purposes, 11*, 129–139.

Hutchins, J. (1977). *On the structure of scientific texts.* UEA papers in Linguistics 5, September, 18–39.

Jamali, H. R., & Nicholas, D. (2010). Intradisciplinary differences in reading behaviour of scientists. *The Electronic Library, 28*, 54–68.

Mason, C. L., Kahle, J. B., & Gardner, A. L. (1991). Draw-a-scientist test: Future implications. *School Science and Mathematics, 91*(5), 193–198.

Mead, M., & Métraux, R. (1957, August 30). Image of the scientist among high-school students. *Science, 126*, 384–390.

Meldrum, A., Boatner, L. A., & Ewing, R. C. (2002). Nanocrystalline zirconia can be amorphized by ion radiation. *Physical Review Letters, 88*, 025503-1–025503-4.

Mourez, M., Kane, R., Mogridge, J., Metallo, S., Deschatelets, P., Sellman, B., et al. (2001). Designing a polyvalent inhibitor of anthrax toxin. *Nature Biotechnology, 19*, 958–961.

Myers, G. (1991). Lexical cohesion and specialized knowledge in science and popular science texts. *Discourse Processes, 14*, 1–26.

Norris, S. P. (1985). The philosophical basis of observation in science and science education. *Journal of Research in Science Teaching, 22*, 817–833.

Norris, S. P. (1992). Practical reasoning in the production of scientific knowledge. In R. A. Duschl & R. J. Hamilton (Eds.), *Philosophy of science, cognitive psychology, and educational theory and practice* (pp. 195–225). Albany: State University of New York Press.

Norris, S. P., & Phillips, L. M. (1994). Interpreting pragmatic meaning when reading popular reports of science. *Journal of Research in Science Teaching, 31*(9), 947–967.

Norris, S. P., & Phillips, L. M. (2003). How literacy in its fundamental sense is central to scientific literacy. *Science Education, 87*, 224–240.

Norris, S. P., & Phillips, L. M. (2008). Reading as inquiry. In R. A. Duschl & R. E. Grandy (Eds.), *Teaching scientific inquiry: Recommendations for research and implementation* (pp. 233–262). Rotterdam: Sense.

Norris, S. P., Falk, H., Federico-Agraso, M., Jimenez-Aleixandre, M. P., Phillips, L. M., & Yarden, A. (2009). Reading science texts – Epistemology, inquiry, authenticity – A rejoinder to Jonathan Osborne. *Research in Science Education, 39*(3), 405–410.

Royal Society. (2013). *Proceedings of the Royal Society B*. Retrieved June 4, 2013, from http://rspb.royalsocietypublishing.org/site/misc/about.xhtml

Song, J., & Kim, K.-S. (1999). How Korean students see scientists: The images of the scientist. *International Journal of Science Education, 21*, 957–977.

Suppe, F. (1998). The structure of a scientific paper. *Philosophy of Science, 65*, 381–405.

Swales, J. (1981). *Aspects of article introductions*. Birmingham: University of Aston.

Swales, J., & Najjar, H. (1987). The writing of research article introductions. *Written Communication, 4*, 175–191.

Tenopir, C., & King, D. W. (2001). The use and value of scientific journals: Past, present and future. *Serials, 14*(2), 113–120.

Tenopir, C., & King, D. W. (2002). Reading behaviour and electronic journals. *Learned Publishing, 15*, 259–265.

Tenopir, C., & King, D. W. (2008). Electronic journals and changes in scholarly article seeking and reading patterns. *D-Lib Magazine, 14*(11/12). Retrieved from http://www.dlib.org/dlib/november08/tenopir/11tenopir.html. 2 Mar 2011.

Wonham, M. J., de-Camino-Beck, T., & Lewis, M. A. (2004). An epidemiological model for West Nile virus: Invasion analysis and control applications. *Proceedings of the Royal Society of London, Series B: Biological Sciences, 271*(1538), 501–507. doi:10.1098/rspb.2003.2608.

Chapter 4
APL and Reading in Science Classrooms

In Chap. 2 we discussed one important reason for using APL to teach science in schools: that is, APL is based upon authentic scientific practice. By "authentic" we mean that APL conforms to the original scientific writings upon which it is established in such a way as to replicate the essential features of those original writings. Because APL has this characteristic, it is able to support authenticity in science instruction, which we have described as "engaging students in posing questions, designing their own paths to solve them, collecting evidence, evaluating claims against evidence, [and] building arguments in a dialogic setting" (Yarden et al. 2009, p. 393). We understand each of these engagements to be part of an authentic scientific practice that is key to learning science and its nature.

Chapter 3 was about the importance of reading by scientists and about the structure of scientific texts. However, the fact that reading is important to scientists does not imply that reading ought to be important to science education. Also, there is no direct implication from the fact that scientific texts are structured in various ways to how science ought to be taught in schools or at other educational levels. Therefore, any claims about the importance of reading to science *education* or about the structure of science texts that students are asked to read need to be defended on educational grounds. Of course, those educational reasons might appeal to the importance of reading in science and to the structure of scientific texts, but a justification must be constructed that links these facts about science to the goals of science education. It is on developing such a justification that the first part of this chapter focuses.

The second part of the chapter will deal with more practical questions concerning how APL works as a tool for teaching reading in science. Specifically, we show how APL can support scientific inquiry, enhance conceptual understanding of science, and promote a deeper grasp of the nature of science. We focus the bulk of our attention on the first of these three topics.

© Springer Science+Business Media Dordrecht 2015
A. Yarden et al., *Adapted Primary Literature*, Innovations in Science Education and Technology 22, DOI 10.1007/978-94-017-9759-7_4

Reading Science as a Proper Goal of Science Education

The justification that we propose is based upon four claims about reading in science:

1. Reading is a central practice of scientific inquiry just as are observation, measurement, and analyzing data;
2. The accepted goal of science education for most students—scientific literacy—must include teaching students to read scientific text;
3. Reading scientific text must be understood as "comprehending, interpreting, analyzing, and critiquing texts" (Norris and Phillips 2003, p. 229) and not as "decoding words and locating information" (Norris and Phillips 2012, p. 49); and
4. Reading scientific text that conforms to the description in claim 3 provides insights into the nature of science, including scientific epistemology.

Reading Scientific Text Is Part of Scientific Inquiry

Science currently is understood to comprise a series of practices. Taken together, these practices define science. A first approximation to understanding what we mean by practices can be had by some examples of what we do not mean. First, scientific practices do not mean the scientific method. Conceptions of scientific method range from simple and restrictive lists of five or six steps to generalities with little power to enlighten, such as, the scientific method is whatever scientists do when they are doing science. Such conceptions do not speak to images that are called to mind by practices, such as professional engagement and proficiency. If we think of medical practices or legal practices as examples, these two elements come frequently to mind, and so they should for scientific practices. Second, scientific practices as we conceive of them are not equivalent to scientific processes, at least as those processes often have been described in science education. For example, scientific observation has been characterized as a simple mental process, in fact the simplest of all scientific processes. To think of scientific observation as generally simple surely is to misrepresent it (Norris 1985). Moreover, to equate observation to a mental process is to individualize and internalize it. In fact, scientific observation involves, as often as not, a cooperative engagement of more than one scientist and involves overt material acts that are readily visible. Third, by practices we do not mean behaviors, especially as understood by behaviorist theory, which, despite being long in the tooth, still survives surreptitiously in much educational thought. Behaviors, as understood by that theory, are totally defined by what is physically manifest. So, for example, just as the full meaning of scientific observation cannot be captured by internal mental processes but must include externalized material acts, its full meaning cannot be represented solely by the physical without reference to the thought behind the physical.

A scientific practice, then, involves a community of practitioners engaged through overt material acts in a domain of mutual interest with common goals.

The practice is upheld by communally shared and developed specific competencies that enable scientists to take part in jointly performed and cooperative activities. There are many such practices. Some practices are broad in scope, such as, developing a research proposal, conducting an experiment, and building a theoretical model. Other practices are narrower in scope and can combine to form the broader practices: observing a phenomenon, collecting and compiling data, designing an apparatus, and analyzing data.

Is reading scientific texts such a practice? More specifically, is the creation (to include writing), interpretation, analysis, and critique of scientific texts by scientists one of the practices that define science and that make up scientific inquiry? We have contended in Chaps. 2 and 3 that the reading scientists do as part of their work has all of the hallmarks of a practice. Here, we formalize our argument for believing this contention to be true. First, reading engages scientists in their professional work for a large proportion of their time. We have documented evidence to support this claim in Chap. 3. Scientific texts are the vehicle both for transmitting new scientific knowledge and for receiving it. Second, the reading scientists do is part of a cooperative engagement aimed towards common goals. The cooperative effort is clear: scientists write for other scientists to read; scientists read to learn what other scientists are thinking as displayed in what they have written. The goal of advancing scientific knowledge is shared by all. Although both writing and reading typically are solitary acts, they are acts performed in the context of a community of practitioners taking part in a joint endeavor. That is, writing and reading scientific texts also are connected to public acts that are constituent elements of scientific inquiry. Third, writing and reading scientific text require acknowledged proficiencies. Much effort both during the formal education of the scientist and during professional life is devoted to honing the proficiencies that are required to write clearly and effectively and to read analytically and critically. Based upon the foregoing reasons, we have confidence in concluding that reading is a central scientific practice and that it is essential to scientific inquiry.

Thus, we conclude that reading, in the same manner as other scientific practices, such as observation, measurement, and analyzing data, has a *prima facie* call for inclusion within the science curriculum that aims to teach scientific inquiry. Of course, this call might be acknowledged while at the same time holding that the claim for inclusion that reading has upon the science curriculum is, in fact, small; that other practices, such as perhaps, observation and measurement, have a much larger claim. We disagree! In the first place, reading typically is required in order to make an observation or a measurement. Furthermore, reading in fact serves at least as much as these other practices as a generative source of scientific ideas. The impetus for making an observation or a measurement often stems from something that has been read—reading is a wellspring of scientific ideas. Thereafter, the outcomes of observations and measurements are put into forms that themselves can be read so that they can serve as evidence for claims and as a source of ideas for others. If reading is of such primary importance to science, then teaching reading in science must be a feature of any science education that takes seriously scientific ideas and the nature of science.

Scientific Literacy in Its Fundamental Sense

Scientific literacy is endorsed broadly as a goal of science education, along with attracting the next generation of scientists into the field. Scientific literacy frequently is considered the more important of these goals for three reasons. The first reason is that scientific literacy is directed primarily towards those students who will not become scientists or have scientific careers, and this group makes up the vast majority of students in school. The second reason is that it is difficult if not impossible during school years to identify those who will become scientists once they finish school. Therefore, those who will become scientists are difficult to target with specific offerings geared to their specific needs. Third, there are reasons to believe that the school science education suitable for most students is, at least at school age, the one also most suitable for future scientists. After all, scientists want to communicate with non-scientists and certainly look to them for support of the scientific enterprise. Thus, it could be useful for scientists to understand how non-scientists' school science education prepared them to think about science. Furthermore, it might be helpful for scientists to step outside their roles as scientists from time to time and to examine science from an outsider's perspective. Scientific literacy is, consequently, an important science education goal for all (Norris and Phillips 2015).

The issue is not completely straightforward, because "literacy" has two main meanings: being educated and cultured, and being able to read and write (Norris and Phillips 2012). The first of these, being educated and cultured, is the meaning generally and traditionally attached to scientific literacy. Thus, one who is scientifically literate has been considered knowledgeable of the substantive content of science, and, because science is a mark of human excellence and achievement, one knowledgeable in science has been considered cultured and accomplished. It is easy to grasp how this view of scientific literacy has been deemed significant. For this reason, and perhaps for other reasons as well, science education has concentrated almost exclusively on imparting the substantive content of scientific ideas, theories, and laws. Some attention has been paid to teaching the nature of science, and some also to teaching about scientific inquiry, but these goals have played lesser roles in science education.

Concentrating on scientific literacy in this manner has steered attention almost completely away from the other main meaning of scientific literacy—the ability to read and write in science. There is no argument that we have found that can justify the neglect of one of the main aspects of scientific literacy and the promotion of only the other. Furthermore, Norris and Phillips (2003) have provided an extensive argument for the conclusion that neglecting reading and writing in science is to give insufficient attention to one of the major constituents of scientific practice, a point related to the one made in the previous section. Therefore, if scientific literacy is to be accepted as the primary goal of science education for most students, and if, as seems the case, there is no justification for neglecting one of the primary senses of scientific literacy at the expense of the other, then the goal of scientific literacy must attend to both meanings of the term. Hence, reading science is a proper goal of science education.

Reading as a goal is not to be treated lightly. Reading is not simply the recognition of words and the location of information, as it often has come to be taken (Collins Block and Pressley 2002; Norris and Phillips 2012; Pressley and Wharton-McDonald 1997). When students hold this simple view of reading, they tend to make a number of predictable errors in their reading of science (Norris and Phillips 1994), the most significant of which is that they overestimate drastically the degree to which they have understood the text (Norris et al. 2003). Sophisticated reading of science must be understood as comprehending, interpreting, analyzing, and critiquing texts, which is the topic of the following section.

Critical Scientific Literacy

Any reading that goes beyond literal reading must be critical. We can substantiate this claim by noting, first, that reading beyond the literal implies making inferences about meaning. Inference always carries with it some level of uncertainty. Dealing with such uncertainty entails the consideration of alternative interpretations, and weighing and balancing the alternatives to determine which interpretation best fits the text (Phillips 1988). The adjudication of alternative interpretations is a very large part of what is understood as critical thinking (Norris and Ennis 1989; Siegel 1988), which establishes the point that reading beyond the literal must be critical.

In the preceding paragraph, we distinguished between the literal interpretation of text, which we understand as the surface meaning or non-inferential meaning of language, and inferential interpretation. In science, inferential interpretations must be made frequently. When reading scientific text, one type of inferential interpretation that must be made is of pragmatic meanings, which involves inferring "meaning by using the context to decide on authors' intentions" (Norris and Phillips 1994, p. 948). Pragmatic meanings are very important in science. For example, if a reader interprets the pragmatic meaning of an author "to be stating something as an unlikely possibility ... then this interpretation can be used to help decide whether the statement is being offered in support of another" (Norris and Phillips 1994, p. 952). Think through a general case. Suppose it is clear that an author has claimed some statement q to be true. Suppose that the author also has made the statement p, but the reader infers that the author believes that p is unlikely to be true. If, to support q, p needs to be true, then the reader is justified in inferring that the author is not offering p in support of q. So, the reader might conclude, the author must be talking about p for some other reason, and then attempt to discern what that reason might be. It is clear that correctly interpreting in the first place the author's pragmatic meaning is vital to understanding other key relationships among statements the author makes.

It is more than two decades ago since a special issue of the *Journal of Research in Science Teaching* published in 1994 recognized the central and important role of reading both in the conduct of science and in the learning of science. Before that time, and even in much research since, scientific language was assumed to be exchanged in what philosophers would call "extensional" contexts. In extensional contexts, the intentions of language users do not need to be probed in order to understand the meaning of the language. Extensional contexts were thought essential to the conduct of science for a number of reasons, including the permissibility of replacing equals with equals in accord with the transitive law. Consider an example. We have taken some liberty with the facts that do not influence the point we wish to make. Take yourself back to the day when ancient astronomers knew of five heavenly wanderers and had named one of them "Venus". Subsequently, a discovery was made through carefully timed observation that Venus is the same as the Morning Star, that brightest of heavenly objects that was known sometimes to accompany the Sun during sunrise. Subsequently still, another discovery was made, namely, that Venus is the same as the Evening Star, that "other" very bright object known to be a companion to the Sun from time to time when the Sun is setting. Lo and Behold! A new scientific discovery had been made, and it was made possible by the extensional context of scientific language: Since Venus is the Morning Star, and Venus is the Evening Star, the astronomers could conclude, substituting equals for equals, that the Morning Star is the Evening Star. This discovery actually did have a profound effect upon our understanding of the Solar System.

Now consider another context. Suppose that Socrates believed that Venus (that wanderer high in the night sky) is the Morning Star (that very bright companion to the Sun during occasional sunrises). Does the fact that the Morning Star is the Evening Star allow us to conclude that Socrates believed that Venus is the Evening Star? All we have done is to replace equals by equals as before. The difference between the examples is that in the case of Socrates the context is not totally extensional. This is so because the context of part of the language use is about what Socrates believed. This context includes Socrates' intentions. Therefore, just because Socrates believed that Venus is the Morning Star and the Morning Star is the Evening Star, does not permit us without further information to infer that Socrates believed that Venus is the Evening Star. We would have to ask Socrates in order to find out his intentions. Perhaps Socrates did not believe that the Morning Star is the Evening Star. In that case, he would not be compelled in order to maintain self-consistency to believe that Venus is the Evening Star.

For many philosophers of science, and indeed for many scientists, it is intolerable to think that in scientific reasoning one could follow the transitive law and the inferences not hold. The outcome of this unbearable consequence was the insistence that the context of scientific language use must be extensional. *Ideally, all properly scientific meanings are literal*, is a shorthand but reasonably accurate and not too facetious way to put this point. Thus, a reported scientific observation should have universal meaning and truth value, that is, meaning and truth value independent of the intentions of the language users reporting the observation.

The switch in science education research to a focus on pragmatic meanings contained in scientific texts was a switch to a focus on scientists' intentions. The upshot is that scientific discourse can require interpretation that takes into account what the writer or speaker intended. This switch carries with it the necessity for scientific literacy to be critical. Olson and Babu (1992, p. 184) maintained that the "interpretation, analysis, and criticism of written texts is what critical thinking is and what it is for." Hence, they argued that critical thinking is synonymous with literacy. The disposition to engage in such criticism arises from the scientific epistemology of fallible rationality discussed in Chap. 3. Just because something is asserted, even by an expert, it is fallible and must, therefore, be questioned (Feynman 1969).

So we see that scientific literacy must be critical. First, as established in the previous section, scientific literacy must include reading science as part of its objective. Second, we have shown in this section that reading science to any degree beyond the literal involves the interpretation of pragmatic meanings, which necessarily involves criticism. This is a very good position in which science education should find itself. If science education is to be properly educative and to play a full and even leading role along with other cultural achievements of humanity in the liberal education of students, then that education cannot be doctrinaire. This is so because the requirements of liberal education demand respect for the rational powers and current and potential autonomy of students (Peters 1973; Siegel 1988). The surest way to afford such respect is to teach students how to think for themselves and to encourage them to be critical in the classroom of the ideas they are being taught. Such a classroom atmosphere is crucial to developing and encouraging the use of the skills needed for socio-scientific decision making.

Reading Science and the Nature of Science

Norris and Phillips (2008) have argued that reading is inquiry. In the science education literature historically, science has been described as an inquiry into the meaning of natural phenomena. The point of Norris and Phillips' argument was to equate reading and science as processes for interpreting meaning. The strategy in the argument was to contrast reading as inquiry with a simple view of reading that dominates much reading instruction in school and that is found widely in science education. As described previously in the section on the fundamental sense of scientific literacy, the simple view of reading is defined by word recognition and information location. According to the simple view, knowing what a text says is equivalent to knowing what the text means (Olson 1994). As we have shown in Chap. 3, reading according to the simple view does not resemble the reading that scientists do in their work. Yet, even a curriculum such as the *BSCS Green Version* (BSCS 1973) founded on science as inquiry, failed to recognize the importance of reading in scientific inquiry when it claimed that "experience in scientific work . . . cannot be gained by reading . . . [but] only by doing the kinds of things scientists do in their laboratories" (pp. viii–ix).

This line of argument is flawed in several ways. First, it assumes that all scientists work in laboratories. Even if we allow for metaphorical meanings, such as field work being performed in nature's laboratory, some scientists simply do not work in laboratories. Rather, they work at their desks, reading and thinking up new ideas and writing them down. Second, it takes for granted that reading is not part of scientific work. We have shown in Chap. 3 that reading indeed is a large part of the work that scientists do, along with writing. Third, and this is a corollary of the second point, it is claimed that reading cannot provide scientific experience. We make the counterclaim based on the evidence presented in Chap. 3 that, if the reading is of scientific papers, then it certainly can provide experience in scientific work.

One of the key aspects of experience that reading scientific papers yields is access to how scientists think. In our Prologue, we quoted Joseph Schwab, who has put this idea well. We repeat the part of the quotation relevant to the current point:

> Papers by scientists reporting scientific research have two major advantages as materials for the teaching of science as enquiry. One advantage is obvious. They afford the most authentic, unretouched specimens of enquiry that we can obtain. (Schwab 1962, p. 81)

Why might Schwab have thought that scientific research papers are the most representative exemplars of scientific inquiry? Is it not observing, measuring, experimenting, and analyzing data that fills that category? Well, Chap. 3 demonstrated how several scientific papers, which we used as instances, did exemplify scientific inquiry. Basically, the papers displayed how the scientists thought. Their reasoning, and more specifically their epistemology, which we called "fallible rationality", was evident. Indeed, scientific reasoning, which is what inquiry concerns, was so much on display that it is not at all unapt to claim that the major purpose of the papers was to record scientific reasoning. Thus, the papers were "the most authentic, unretouched specimens of inquiry" for the very reason that their major purpose was to provide authentic, firsthand official records of the scientific inquiries in which the scientists were occupying themselves and of which the papers formed a part.

The question that we asked in the introduction to this chapter now arises naturally: Why, given our analysis of the connections between reading and science, is it a proper goal of science education to teach students how to read papers such as these and to make it a requirement of the curriculum that they do read them? The answer harks back to the reasons why it is important to teach students to think critically about science. Scientific papers record the inquiry process. As such, they demonstrate the source of scientific conclusions and provide the grounds upon which those conclusions are based. Teaching students the provenance of and grounding for knowledge provides for them the first point of entry into the critical appraisal of that knowledge. This educational approach shows respect for students as human beings, and recognizes their possession of rational abilities and potential for autonomous action. This justification is one of the strongest that can be provided for an educational effort because of its recognition of students as ends in themselves rather than as means, as citizens whose fulfillment and autonomy are important goals in and of themselves.

APL as a Tool to Teach Reading in Science

In this section of the chapter we concentrate on how APL can be used to inspire and support inquiry. We also comment on how APL can be used to teach for conceptual understanding and to develop knowledge and insight into the nature of science. In helping to accomplish these goals, APL use supports the justifications for teaching reading in science examined in the preceding section.

APL Can Inspire and Support Inquiry

We have made it clear how reading is part of the inquiry process of science, due to the structure of primary scientific literature (PSL) and the interpretive demands that it makes. PSL authors ask questions, provide answers with supporting reasons and evidence, consider alternative answers and attempt to impugn those they judge unsuitable. Casting each of these activities in the form of written words is an element of inquiry, as is subsequently reading, interpreting, and critiquing those words. A serious consideration of using APL for school science instruction brings us back to Schwab's exhortation. Recall that he saw scientific papers as "the most authentic, unretouched specimens of enquiry that we can obtain" (Schwab 1962, p. 81). APL is not "unretouched", but to the extent that it maintains the canonical form of the originals, it should nevertheless provide authentic specimens of inquiry in the sense that Schwab intended.

So, we need to imagine the consequences of using authentic specimens of inquiry to teach school science. It certainly is reasonable to conjecture that APL would support inquiry by students. However, reliance on conjecture is not needed because we have evidence that APL use supports students' scientific inquiry. Similar studies were conducted in Israel and in Canada (Baram-Tsabari and Yarden 2005; Norris et al. 2012). In both studies, high school students were randomly assigned to read one or the other of two adaptations of a published scientific article, one written in APL format and the other written as a popular scientific genre (Journalistic Reported Version, JRV). In both studies, students were taught relevant background material in prior classes, but they had not seen the readings before they were asked to read them and answer sets of questions. The questions related to their interest in the articles, their comprehension of what they had read, and the critical stance they adopted towards the articles. The results most pertinent to our discussion here are that in both studies students who had read the APL displayed a more critical stance toward what they had read. Compared to the JRV readers, the APL readers listed more research that could be conducted in order to further test the accuracy and generality of the findings, and cited more scientific criticisms of the researchers' work by pointing to more parts of the research that could have been accomplished differently and by supplying a greater number of applications of the technique that was used by the researchers.

The APL readers behaved differently from the comparison groups in these studies but also vastly differently from students in earlier studies who had read secondary literature. For example, in a study by Phillips and Norris (1999), students asked to read secondary literature were found to defer "to the reports by readily accepting the statements of the reports and by implicitly trusting the authors. On only rare occasions did readers challenge the authority of the reports or the authors" (p. 325). Indeed, it was found that there was no systematic relationship between what students believed before they read the reports, what they had read, and what they believed after reading. Most typically, they took the reports to supply evidence to support their prior beliefs, even if the reports were irrelevant to their beliefs, and, most shockingly, even if the reports contradicted their beliefs. Thus, when the students seized upon supportive reports as support for their beliefs, the authors expressed little confidence in this consistency because of the widespread inconsistency in their reasoning otherwise.

So why might students reading APL display a more critical response to the text? Our conjecture is that, because the APL texts display inquiry, they also invite inquiry. We believe that the invitation might be conveyed via several mechanisms. First, the APL can serve as a model of good scientific writing and thinking. The students must notice that the APL format is different from anything else they have read, unless they have been reading PSL, which is not likely. A little reflection would reveal that a major difference is that, compared to their science textbooks, APL engages in far less enumeration of facts to accept and in far more provision of arguments meant to persuade rationally. For this reason, students might take the APL approach as far more respectful of their intelligence and thus as something to emulate. Second, APL might be seen as providing a license to do what normally is not expected in school. By being argumentative, APL could be interpreted as an authorization to argue, especially when accompanied by the implicit endorsement of the science teacher who is asking students to read it. This mechanism can work in tandem with the modeling one—APL both demonstrates what good scientific thinking is like and sanctions that mode of thought. Third, APL might simply inspire. APL requires students to use their brains in ways that are novel for school. We remember clearly the student who exclaimed when reading an APL text, almost as if he were learning a secret about science, "Is that what scientists really do?" He had just finished reading the section of an APL that listed the various limitations and idealizations that Wonham et al. (2004) had imposed when building their model of the spread of the West Nile virus. The student's astonishment seemed to have two sources: (a) he was amazed that a published piece of scientific research could have so many places where it could fail, and (b) he was further amazed that the authors would admit openly to these limitations. Such a realization can be inspirational, mainly because of the sense of liberation that attends it. Perhaps for the first time the student saw it as scientific to admit to an imperfection, to a defect that could undermine the work. Realizing science has this characteristic might be sufficient to make it seem less intimidating, even more human.

In each of the following two sections, we focus on scientific articles we have discussed in Chap. 3, and a pair of adaptations of each article, one adaptation written as an APL and the other as a JRV popularization. Our point is to illustrate through the comparison of features how APL can inspire and support inquiry. In the first of the sections, we examine adaptations of the article on designing an inhibitor of the anthrax toxin (Mourez et al. 2001) from the point of view of how inquiry is represented through their structure. In the second section, we appraise the differences between two adaptations of the article providing an epidemiological model of the West Nile virus (Wonham et al. 2004) with respect to their exhibition of tentativeness.

The Various Ways That Inquiry Is Represented in APL

In order to follow this section you will need to read the two adaptations of the anthrax article found in Chap. 8. We will draw upon the three meanings of structure that we introduced in Chap. 3: organizational, goal-directed, and argumentative.

First, let us compare the structures of the two adaptations at the organizational level. Compare the titles. The title of the JRV version, with its use of the description "A Successful Experiment", is unlikely to be found in the title to a published scientific article. The JRV title suggests with its use of the word "successful" that inquiry into the questions raised by the experiment is closed. So, from the very beginning, the JRV piece tends to foreclose further inquiry into the topic. Note that there are no section breaks in the JRV article. In contrast, the APL article is divided into four sections. The section titles, with the possible exception of "Introduction", are indicative of elements of inquiry: Materials and Methods, Results, and Discussion. The suggestion in all of these headings is that an inquiry is being reported. An important distinction between the JRV and APL articles is that the latter separates the description of the materials and methods from the reporting of results and from the discussion of the results. The APL version also devotes more words to these sections. The JRV version not only lacks headings to keep these topics separate, the topics often are discussed together in a single paragraph. For instance, consider the following paragraph. The first sentence describes purpose and method. The second sentence reports a result before the semi-colon. After the semi-colon, the phrase describes an additional method before the comma, and reports a result after the comma.

> In order to find out whether the peptide effectually inhibits the assembly of the toxin in living cells, the researchers added the heptamer, the LF protein and the inhibiting peptide to hamster ovary cells grown in culture. They found out that the peptide conveyed a mild inhibiting activity to the binding of the LF enzyme and the cleaved PA protein; adding a different peptide that served as a control, did not yield a similar effect. (see Chap. 8)

Separating topics using headings as done in the APL version is significant from the perspective of representing inquiry. The significance lies in the epistemic status of the topics. The methods and materials and the results sections both contain reports: in the former, a report of what was done; in the latter, a report of what was found. The fact that they are reports is significant for determining the nature of the evaluation and critique suitable to them. Typically, a critique of the methods and results section would not claim that the scientists provided an inaccurate report of what was done. (Of course, we can imagine unusual circumstances in which this indeed was the substance of the critique.) Rather, a critique typically would be about what *should have been done*, pointing out a better, more efficient, or more exact procedure. Similarly, in normal circumstances, a critical review of the results would not claim that the scientists did not find what they reported finding, in the sense of claiming that these were not the data collected from their instruments. More likely, a criticism would be about whether these results are probable given the materials and methods, whether there likely was some defect in the instrumentation, or whether the scientific technique must have been flawed and led to results not to be expected. The mistakes, thus, are not suspected to be found within the reporting of what was done or data collected, but rather within the choices of materials and methods and within the instrumentation and technique that yielded the results reported. The paragraph above from the JRV version also contains a brief description of purpose, which has an epistemic status different from both the methods and materials and the results descriptions. The purpose has to be evaluated using context-specific value reasoning: Is the goal of the study sufficiently valuable to be worth the time, expense, and effort?

A critique of the discussion section can adopt yet another approach. The discussion section provides interpretations of the results and of their significance for theory and application. A challenge to these interpretations must provide reasons why a given interpretation is mistaken, and, usually, an alternative interpretation of the results that overcomes the reputed flaws of the one being criticized. Similarly, a challenge to the claimed significance of the results must target those claims with reasons meant to undermine them. The difference between the two approaches to critique is that the materials, methods, and results can be challenged without making any commitment to the interpretations found in the discussion. However, a challenge to the interpretations must accept, at least for the nonce that the results are as reported.

Given the different strategies and grounds for criticism depending upon the topic being examined, it is much more in conformity with the nature of inquiry to assemble into one section the purpose of the study, into another section all of the reports of background information, into another the materials and methods descriptions, into another the results, and into another the discussion. This organizational technique represents how the scientific inquiry is conceived from an epistemic point of view. The topics are arranged according to how a potential reader could approach developing a criticism.

Next, consider the goal-directedness of both versions. In the APL version, the primary goal was announced at the end of the third paragraph: "Thus, developing a specific inhibitor to the toxin might yield a beneficial addition to the use of antibiotics." The major goal of the paper is more specifically detailed at the beginning of the sixth paragraph: "This article will describe the development and testing of an inhibitor of anthrax toxin that works by binding to the PA protein heptamer and preventing its binding to the LF enzyme." Examples of subsidiary goals often linked to methods appear in various places throughout the document.

1. Here are examples of goals from the materials and methods section—

    The search started with finding a phage that binds to the cleaved and activated PA protein in the vicinity of the site where the PA naturally binds with the LF enzyme so it could interfere with their interaction . . .

    After an incubation period, the surface was washed in order to discard phages that do not express peptides that bind to the PA protein.

    In order to confirm molecular binding, molecules were tagged with radioactive markers.

    In order to estimate the ability of the inhibitor to inhibit the biological activity of anthrax toxin, the ovarian cells were incubated with PA and LF proteins that carry a portion of the diphtheria toxin.

2. Here are examples of goals from the Results section—

    In order to inhibit the activity of anthrax toxin, a protein that interferes with the assembly of its constituents was searched for . . .

    The next stage was to prove that the inhibitor effectively inhibits the biological activity of the toxin.

    The diphtheria toxin part is aimed to show whether the proteins are able to enter the cells or not . . .

    The last experiment, aimed at proving the therapeutic potential of the inhibitor

3. Here are examples of goals from the Discussion section—

    The peptide was then used in order to produce a more efficient inhibitor . . .

    The efficacy of the inhibitor in preventing the activity of the anthrax toxin in animal models suggests that it, or another inhibitor developed by a similar approach, may serve as a therapeutic agent for patients infected with anthrax.

There are expressions of goal-directedness within the JRV version as well. Consider the following examples:

Attempting to prevent the lethal activity of the anthrax toxin, the researchers decided to try to interrupt with the assembly of its subunits. The idea was to find a protein that binds the cleaved and active PA protein . . .

In order to find out whether the peptide effactually inhibits the assembly of the toxin in living cells, the researchers added the heptamer, the LF protein and the inhibiting peptide to hamster ovary cells grown in culture.

To assess the effect of the new inhibitor comparing to the single peptide, they [the researchers] again added each of them, separately, to hamster ovary cells, together with the constituents of the toxin.

Next, the researchers attempted to assess the activity of the synthesized inhibitor in preventing intoxication.

First, we note that the examples chosen from the JRV version are exhaustive, whereas the ones from the APL version represent a selection from a larger pool of examples. Thus, there was a greater sense of goal-directedness in the APL version. Given that goal-directedness is a structural feature of the examples of primary scientific literature we have canvassed, the APL version more closely resembles these original reports of inquiry on this characteristic.

There is another difference that has to do with the relationship between the subjects of the sentences and the actions being described. The statements of goal-directness expressed in the APL version are written mostly in the passive voice, but they are written in the voice of the researchers. For example, the first example under the Results section says, "a protein that interferes with the assembly of its constituents was searched for". Converted into active voice, the researchers could easily have said, "We searched for a protein that interferes with the assembly of its constituents." In contrast, the examples from the JRV version are written mostly in active voice. Consider the first example, "the researchers decided to try to interrupt with the assembly of its subunits". An initial temptation might be to associate the active voice with inquiry more than the passive. Yet, it is the narrator of the JRV and not the researchers who is doing the speaking in the JRV example. The use of the active voice, compared to the passive in the APL, might lead to the conclusion that the JRV is more goal-directed. However, the effect of the active voice is outweighed by the difference in speaker: in the JRV it is a narrator talking about the researchers; in the APL it is the researchers themselves who do the talking. The fact that it is the researchers themselves who speak and the fact that they talk about their goals more frequently combine to give the APL version a greater sense of goal-directness.

We next turn to the argumentative structure of both articles. Recall from Chap. 3 that by argumentative structure we mean the logic of the reasoning contained in the text. Is there data offered to confirm or disconfirm hypotheses? Is there a theoretical argument for a conclusion? Are predictions made? Consider first the JRV article. Here are some examples of the type of reasoning we are looking for:

When they added the lethal portion of the LF protein to the solution, it interrupted with the binding of the peptides to the heptamer. Hence, the researchers concluded that the peptides bind the PA protein at the same site where the toxic enzymes are supposed to be bound ...

In order to find out whether the peptide effectually inhibits the assembly of the toxin in living cells, the researchers added the heptamer, the LF protein and the inhibiting peptide to hamster ovary cells grown in culture. They found out that the peptide conveyed a mild inhibiting activity to the binding of the LF enzyme and the cleaved PA protein; adding a different peptide that served as a control, did not yield a similar effect ...

To assess the effect of the new inhibitor comparing to the single peptide, they again added each of them, separately, to hamster ovary cells, together with the constituents of the toxin. Assembly was assessed through radioactive protein tagging (Figure 8.5a, p. 158).

The experiment demonstrated that the inhibitor interfered with the binding of the PA protein with the LF enzyme much more efficiently and in much smaller concentrations. In fact, the activity of the inhibitor was 7,500-folds greater than that of the single peptide. A control substance (the flexible backbone without the peptides) had absolutely no interfering effect on the assembly of the toxin.

Next, the researchers attempted to assess the activity of the synthesized inhibitor in preventing intoxication. ... As shown in the graph (Figure 8.5b, p. 158), the inhibitor prevented intoxication, while the control substance (the flexible backbone alone) had no effect at all ...

The rats were injected with a comparatively large dose of the PA and LF proteins that are responsible together to the lethal symptoms of anthrax disease, and the therapeutic potential of the inhibitor was assessed (Table 8.2, p. 159). ... In the rat model, injection of the inhibitor together with the toxin delayed the symptoms of anthrax from appearing. A higher dose of the inhibitor prevented the symptoms of intoxication completely ... (see the full JRV text on pp. 155–159).

You are able to see that there are several examples of reasoning from evidence in the JRV article. Next, consider the APL article. Here are examples we found:

A few characteristics of the anthrax bacteria make it suitable for use in biological warfare: they can be spread through aerosol (a suspension of microscopic droplets that contain the bacteria); the number of disease-causing bacteria is comparatively small (8,000–10,000 bacteria); the resulting disease carries a high death rate; the bacteria can transform into spores, resistant to being carried in a missile and easy to be spread by the wind; last, the bacteria are easily found in soil ...

In order to confirm molecular binding, molecules were tagged with radioactive markers.

In order to estimate the ability of the inhibitor to inhibit the biological activity of anthrax toxin, the ovarian cells were incubated with PA and LF proteins that carry a portion of the diphtheria toxin ...

When the LF enzyme was added to the solution, the binding of these two peptides was inhibited. This observation supports the hypothesis that the peptides bind the PA protein at the same site as the LF enzyme does ...

Further analysis revealed that both peptides contain an identical sequence of four hydrophobic amino acids. It is thus likely to assume that these amino acids are responsible for binding the heptamer ...

The next stage was to prove that the inhibitor effectively inhibits the biological activity of the toxin. ... Adding the inhibitor protein to the growth media prevented the activity of the diphtheria toxin, while the control substance (the naked flexible backbone) or the single peptide did not affect its toxicity (Figure 8.3b, p. 152).

The last experiment, aimed at proving the therapeutic potential of the inhibitor, was performed on live rats, injected with a high dose of the PA and LF proteins, which are responsible together to the lethal symptoms of anthrax (Table 8.1, p. 153). Injection of the inhibitor along with the toxin proteins effectively delayed the onset of symptoms in rats. Escalating doses of the inhibitor could eliminate the symptoms completely. The rats were also protected when the inhibitor was injected 3–4 min after the toxin proteins. The control substance and the single peptide did not affect the symptoms caused to the rats by the toxin. A 1 week follow-up of the rats did not reveal toxic side-effects for the inhibitor itself ...

The peptide was then used in order to produce a more efficient inhibitor; its efficacy was confirmed both in cell culture and animal models. In cell cultures, it successfully prevented the assembly of the toxin subunits and its entrance to the cells. While injecting the toxin subunits to sensitive rats, co-injection of the inhibitor in small doses (amounts) could delay the onset of symptoms and even completely eliminate them in high doses ...

The efficacy of the inhibitor in preventing the activity of the anthrax toxin in animal models suggests that it, or another inhibitor developed by a similar approach, may serve as a therapeutic agent for patients infected with anthrax (see the full APL text on pp. 146–154).

We have identified several examples of argumentation in both the JRV and the APL articles. The issue of speaker arises once more in these examples. In the JRV version there is a narrator who reports what the scientists concluded or wanted to conclude and the evidence that they found supporting their conclusion. In the APL version the speaker is the scientists reporting what they concluded and what evidence they found to support their conclusions. The difference is subtle, but important for how inquiry is represented in the two modes.

A more significant difference between the JRV and APL versions is that argumentation occurs in more places in the APL article. The JRV does not have explicitly divided sections, but that aside, all of the argumentation in the JRV version is associated with providing evidence from the results for what was concluded. In the APL version, similar argumentation occurs in the results section. However, the APL article also contains argumentation in the introduction, in the material and methods section, and in the discussion section. In the introduction, several reasons are given to support one of the motivations for doing the research, which was that anthrax could be used as a biological weapon. These reasons are in the first example we provided above. In the materials and methods section, the researchers provide two brief arguments for two of their methods—tagging molecules with radioactive markers in order to confirm bonding; and incubating ovarian cells with PA and LF proteins carrying a portion of the diphtheria virus in order to estimate the ability of the inhibitor of the anthrax toxin. In the discussion section, the researchers provide an argument for thinking that the results of the experiment provide hope in finding a therapy for anthrax infections. The existence of argumentation in every section of the APL paper demonstrates that the rationality of science extends to all aspects of inquiry, and is not a feature isolated to defending the conclusions of research.

The Representation and Significance of Tentativeness in APL

In order to follow this section you will need to have read the two adaptations of the West Nile virus article found in Chap. 9. We will build upon the discussion of tentativeness found in Chap. 3. In Chap. 3, we showed that the scientific papers we examined were based upon an epistemology characterized by rationality and fallibility. The recognition of fallibility, we said, evokes a tentativeness in claims. The connection between fallibility and tentativeness is straightforward, because accepting that one is liable to error is bound to make a person wishing to avoid fraud and deception more hesitant, uncertain, and circumspect about drawing conclusions. Such tentativeness is evidenced in scientific papers by the widespread use of language to signal that conclusions are offered without complete certainty. Such language includes the words: approximately, assume, likely, might, perhaps, probably, seems, tentatively, uncertain, and unknown.

Although the APL version of the West Nile virus article proved to contain a greater number of instances of words expressing tentativeness than the JRV version, both contained a large number of instances. The most frequently used works in the JRV version were "perhaps" and "seems", used in the context of colloquial writing. The most common words used in the APL version were "probably", "assumption", and "approximate", deployed in the context of more technical writing. The latter, in our judgment, provided a more reasoned sense of tentativeness, that is, a sense of tentativeness that followed from the nature of the research, which was portrayed more authentically in the APL version. In the following paragraphs we ask you to consider several examples.

Here are examples of the use of "perhaps" in the JRV version. In contemplating how mathematical modeling could help understand the spread of the West Nile virus in the light of attempts to control the spread with spraying and other measures, the authors opine:

> Since all of these strategies would cause additional impacts on the environment, perhaps modelling could help maximize control effectiveness while minimizing unwanted effects? (p. 2)

The usage properly reflects tentativeness regarding the possibility of the usefulness of mathematical modeling. The stance is understandable given that it is written from the perspective of thinking before the modeling had been done. The question of whether a piece of research is worth doing is one that faces scientists all of the time. It demonstrates that tentativeness pervades science from the very outset of even considering whether to conduct a particular study.

In the second example, the context is the early stages of the research when the authors were attempting to learn the relevant characteristics of mosquitoes and their susceptibility to the West Nile virus.

> It can also take quite a long time, perhaps 10–12 days, for an infected mosquito to develop a viral load high enough to transmit the disease back to a bird. (p. 4)

The uncertainty being expressed concerns the amount of a mosquito's life that had to be taken into account by the model. The example shows that the conduct of a study can demand knowledge of the many complexly interrelated parts of nature that bear upon the phenomenon of interest. The difficult question is how to establish a boundary that contains the focal phenomenon and enough of the rest of nature to, on the one hand, make the study practically relevant, and, on the other hand, permit the study to be done within available constraints of talent, time, money, and knowledge. The wider the boundary, the more pressure arises from each of those constraints.

The third and final example, is taken from near the end of the paper at the point where the authors are imagining applications of their findings to localities that have not yet experienced a West Nile virus infection.

> Perhaps the experience of cities and regions with the virus (and therefore with the data) can inform the control programs in areas where a virus outbreak has not yet occurred. (p. 5)

The motivation for the tentativeness expressed here is likely the full knowledge that models such as the one developed cannot be applied mechanistically to contexts different from the one on which the model was based. Application of scientific results to other contexts always requires specific knowledge of those contexts that then needs to be coordinated with the research-based knowledge.

We turn now to examples from the APL version based upon use of the term "assume". Here are a series of examples:

> Because we restricted our model to a single summer, we assumed that any births that year
> had already taken place earlier in the spring, and that the natural bird death rate was
> negligible over that time period ... (p. 3)
>
> We also made some simplifying assumptions to keep the model as straightforward as
> possible. (p. 3) [After this lead-in sentence follow eight statements of assumptions, seven of
> which contain the phrase, "We assumed ...".]
>
> Since we assumed, for simplicity, that the mosquito population remains constant over
> the summer, we set the daily per capita mosquito birth rate, abbreviated b, equal to the death
> rate, k. (p. 4)

All of these acknowledgements of assumptions are directly tied in the article to the
methodology of the study. It is clear the authors are laying them out for the world to
see. If it was not already apparent earlier to the reader, it becomes obvious in the
conclusion section of the paper that the authors recognize that the assumptions they
made bear upon the accuracy that they rightfully can claim for predictions from
their model. They say:

> In future work, we would like to build a series of different models for West Nile virus to
> determine which of our assumptions and restrictions make the model generate less or more
> accurate predictions. (p. 6)

Even if this future work were carried out, it is implicit in the authors' choice of
language that the results would still be tentative, perhaps generating "more accurate
predictions" but never producing perfectly accurate ones.

The authors also represent tentativeness in their frequent use of the word
"approximate" when describing what they know about the lives of mosquitoes
and crows. Repeatedly, they report that knowledge as approximate:

> A female mosquito's lifespan is approximately 33 days, during which time she may lay
> approximately 10 egg masses, each of which takes approximately 15 days to mature into
> biting adults. (p. 3)
>
> Laboratory work has shown that virus transmission occurs at the extremely low level of
> approximately 0.1 %. (p. 3)
>
> Mosquitoes give birth to aquatic larvae, which require approximately 15 days to mature
> into biting adults. (p. 3)
>
> Laboratory work shows that a mosquito that contracts West Nile must incubate the virus
> for approximately 8–12 days before it can be transmitted to a bird. (p. 3)
>
> An infectious crow dies from West Nile virus in approximately 5 days. (p. 3)
>
> It takes approximately 2.3 days for a mosquito to convert a blood meal into a batch of
> eggs, so the maximum biting rate for an individual mosquito is approximately once every
> 2.3 days. (p. 4)

There are many more examples of the same type of usage in which a parameter or
parameters of the system are qualified as being approximate. The acknowledgements are very significant, because they convey to the reader that the basic
knowledge that is being used to construct the model is itself uncertain. The clear
implication is that the model can be no more certain than the parametric knowledge
upon which it is based.

Finally, The APL article makes frequent use of the concept of probability, often
associated with the idea of approximation. Examples include the probabilities of
disease transmission (p. 3, 4) from birds to mosquitoes and from mosquitoes to

birds, and the approximate probability of a crow (p. 3) and of a mosquito (p. 4) dying each day. The frequent use of the probability concept emphasizes the fact that the occurrence of certain events in the space of these phenomena is not deterministic, and the additional frequent coupling of probability with approximation stresses the additional indeterminacy of the knowledge of even the probabilities of these events.

APL and Conceptual Understanding of Science

Genuine conceptual understanding is tied to knowledge of the source and grounding for scientific ideas. Wineburg (1997, p. 259) once said the following about historical knowledge:

> The claim to "know history" is a statement that one knows why an event is important, how it links to other events, what its antecedents were, and how it affected future events. Knowing any one of these aspects in isolation misses the point.

The key to Wineburg's point is that the connectivity of ideas, the linkage from one thought to another, is what makes any single idea comprehensible. The same point can be made about science. Knowing a scientific statement is much more than being able to quote the statement and assent to its veracity. Knowledge of a statement must call upon knowledge of the process or likely process through which the statement was conceived and established, of the degree of certainty that the field attaches to the statement, of the role in reasoning the statement plays in relation with other scientific statements, and of the implications of the statement's being true.

Returning for yet another time to Schwab's claim that scientific papers are the most authentic specimens of scientific inquiry available, we make the claim that APL can help promote just the sort of interrelated knowledge identified in the previous paragraph. APL describes the process through which scientific conclusions are both conceived and established. APL emphasizes the level of certainty with which scientific claims can be advanced. APL demonstrates how scientific statements are interconnected by displaying how reasoning connects one statement to another. Finally, APL traces the implications of scientific conclusions both for the rest of scientific knowledge and for applications beyond science itself. In other words, APL wears on its face the deep conceptual structure of science. We do not expect that these features will automatically leap off the page or screen and make themselves apparent for students. However, we do believe that the use of APL makes it possible for relatively straightforward pedagogical approaches, such as pointing out textual features to students, to have profound effects on students' conceptual understanding of science. Simple pedagogies, such as indicating to students that here the author is reviewing previous research and there introducing the present research, here justifying the relevance of observations and the interpretations of data and there dealing with alternative interpretations, can be effective because the APL provides such a realistic and authentic example. Such instructional directness is not available when using typical science textbooks for the simple reason that the books do not represent science in anything like the manner that Schwab envisioned.

APL and the Nature of Science

We can, yet again, reflect on what Schwab said about scientific text. Rather than focus upon the way such text reflects scientific inquiry we shall concentrate our attention on two other features of APL that bear upon the nature of science. The first feature is their argumentative structure, which we already have discussed at length as a marker for scientific inquiry. There is something general to learn about the nature of science from the argumentative structure of APL. APL provides for students the opportunity to incorporate as part of their image of science the argumentation schema. Schema-based learning is very powerful, as we learned from considerable research in the 1970s and 1980s on how dependent humans are for their interpretations of their environments on fixed and general patterns. Many of us have firsthand knowledge of how difficult it can be to alter stereotypes, which are quintessential schemas. APL makes available the opportunity to capitalize upon the beneficial side of schema-based learning by providing myriad examples of how argumentation is manifested within science.

The second feature pertains to the image of scientists implied by the existence of scientific texts. We touched upon this feature in Chap. 3 when we contrasted patterns of learning and of not learning that are emphasized in much of science education. One of these patterns contrasts the scientist as a laboratory-based worker to the scientist as the creator and interpreter of scientific texts. Evidence has demonstrated convincingly that it is predominantly the former image that students take away from school science education. APL can broaden students' outlooks of scientists to see them also as people who read and who write.

Summary and Conclusion

We have argued in this chapter that reading scientific text is a proper goal of science education. Our justification was based upon four claims to the effect that reading is an essential scientific practice, that scientific literacy must include ability to read scientific text, that reading must be understood as critical reading, and that reading scientific texts provides insight into the nature of science. The upshot of these claims is that reading scientific text provides access to the provenance and vulnerability of scientific knowledge, which lays the foundation for critical scientific literacy—the most important goal of science education. Having established the importance of reading scientific text, we turned to the task of showing that APL is a powerfully effective tool for teaching reading in science. APL has the strong potential to inspire and support scientific inquiry, to foster greater conceptual understanding of science, and to promote knowledge of the nature of science. Although similar claims can be made for other instructional approaches, none of the alternatives is founded on such authentic specimens of inquiry.

References

Baram-Tsabari, A., & Yarden, A. (2005). Text genre as a factor in the formation of scientific literacy. *Journal of Research in Science Teaching, 42*, 403–428.

Biological Sciences Curriculum Study. (1973). *BSCS green version* (3rd ed.). Chicago: Rand McNally.

Collins Block, C., & Pressley, M. (2002). *Comprehension instruction: Research-based best practices.* New York: Guilford.

Feynman, R. P. (1969). What is science. *The Physics Teacher, 7*, 313–320.

Mourez, M., Kane, R., Mogridge, J., Metallo, S., Deschatelets, P., Sellman, B., et al. (2001). Designing a polyvalent inhibitor of anthrax toxin. *Nature Biotechnology, 19*, 958–961.

Norris, S. P. (1985). The philosophical basis of observation in science and science education. *Journal of Research in Science Teaching, 22*(9), 817–833. doi:10.1002/tea.3660220905.

Norris, S. P., & Ennis, R. H. (1989). *Evaluating critical thinking.* Pacific Grove: Midwest Publications.

Norris, S. P., & Phillips, L. M. (1994). Interpreting pragmatic meaning when reading popular reports of science. *Journal of Research in Science Teaching, 31*, 947–967.

Norris, S. P., & Phillips, L. M. (2003). How literacy in its fundamental sense is central to scientific literacy. *Science Education, 87*, 224–240.

Norris, S. P., & Phillips, L. M. (2008). Reading as inquiry. In R. A. Duschl & R. E. Grandy (Eds.), *Teaching scientific inquiry: Recommendations for research and implementation* (pp. 233–262). Rotterdam: Sense.

Norris, S. P., & Phillips, L. M. (2012). Reading science: How a naive view of reading hinders so much else. In A. Zohar & Y. J. Dori (Eds.), *Metacognition in science education: Trends in current research* (pp. 37–56). Dordrecht: Springer.

Norris, S. P., & Phillips, L. M. (2015). Scientific literacy: Its relationship to literacy. In R. Gunstone (Ed.), *Encyclopedia of science education.* Dordrecht: Springer.

Norris, S. P., Phillips, L. M., & Korpan, C. A. (2003). University students' interpretation of media reports of science and its relationship to background knowledge, interest, and reading difficulty. *Public Understanding of Science, 12*, 123–145.

Norris, S. P., Stelnicki, N., & de Vries, G. (2012). Teaching mathematical biology in high school using adapted primary literature. *Research in Science Education, 42*, 633–649.

Olson, D. R. (1994). *The world on paper.* Cambridge: Cambridge University Press.

Olson, D. R., & Babu, N. (1992). Critical thinking as critical discourse. In S. P. Norris (Ed.), *The generalizability of critical thinking* (pp. 181–197). New York: Teachers College Press.

Peters, R. S. (1973). *The philosophy of education.* Oxford: Oxford University Press.

Phillips, L. M. (1988). Young readers' inference strategies in reading comprehension. *Cognition and Instruction, 5*, 193–222.

Phillips, L. M., & Norris, S. P. (1999). Interpreting popular reports of science: What happens when the reader's world meets the world on paper? *International Journal of Science Education, 21*, 317–327.

Pressley, M., & Wharton-McDonald, R. (1997). Skilled comprehension and its development through instruction. *School Psychology Review, 26*, 448–467.

Schwab, J. J. (1962). The teaching of science as enquiry. In J. J. Schwab & P. F. Brandwein (Eds.), *The teaching of science*. Cambridge: Harvard University Press.

Siegel, H. (1988). *Educating reason*. New York: Routledge.

Wineburg, S. (1997). Beyond "breadth and depth": Subject matter knowledge and assessment. *Theory Into Practice, 36*(4), 255–261.

Wonham, M. J., de-Camino-Beck, T., & Lewis, M. A. (2004). An epidemiological model for West Nile virus: Invasion analysis and control applications. *Proceedings of the Royal Society of London. Series B: Biological Sciences, 271*(1538), 501–507. doi:10.1098/rspb.2003.2608.

Yarden, A., Falk, H., Federico-Agrasso, M., Jiménez-Aleixandre, M. P., Norris, S. P., & Phillips, L. M. (2009). Supporting teaching and learning using authentic scientific texts: A rejoinder to Danielle J. Ford. *Research in Science Education, 39*, 391–395.

Part II
The Practice of Adapted Primary Literature

Chapter 5
Creating and Using Adapted Primary Literature

In Chap. 2 we outlined the attributes characterizing four genres of scientific texts used for science learning: Primary Scientific Literature (PSL), Adapted Primary Literature (APL), Journalistic Reported Version (JRV), and textbooks. We discussed the resemblance between PSL and APL in terms of genre, content, organizational structure, and the presentation of science. In this chapter we describe the various steps taken for developing APL, starting from selecting the appropriate PSL along with the various modifications that are incorporated into the text to make it understandable and usable by the target audience. Subsequently, we outline several instructional approaches that were developed for the use of APL in schools, and the benefits and limitations of those approaches. The last part of this chapter focuses on the materials that were developed to support teachers in using APL, including workshops for teachers implementing APL-based curricula and a multi-media teachers' guide with videotaped teaching episodes.

A Step-By-Step Description of How to Develop APL

The various sections of PSL have different rhetorical roles (Swales 2001). The Title and Abstract are aimed at providing a concise presentation of the specific research and the Introduction connects that research with previous experiments, in the context of the general-knowledge domain. The Results section describes the process of manipulating the data obtained using the methods described in the Methods section and makes limited claims about the findings. Finally, the Discussion section explains the findings, raises claims about them, and connects them back to the general-knowledge domain. In APL articles, pedagogical considerations promote the significant conservation of each section's original role and knowledge level during the adaptation process. We describe below the various criteria and the steps taken for choosing a suitable PSL article for adaptation to APL, and the adapting

© Springer Science+Business Media Dordrecht 2015
A. Yarden et al., *Adapted Primary Literature*, Innovations in Science
Education and Technology 22, DOI 10.1007/978-94-017-9759-7_5

process of each section of a PSL article. Finally, we outline the sequence of steps that we usually take while adapting a research article.

Choosing a Suitable Article

The first step in the development of an APL article is to choose a scientific article which is suitable for adaptation. This step proved to be the most crucial step in the development process, since it determines the essence of the text that will be developed from it. Therefore, even though this step is time consuming, it should be carried out with extreme caution and care to ensure the suitability of the article for adaptation.

Three main sources were used for searching suitable PSL articles for adaptation: (i) Digital databases (e.g., PubMed) using keywords; (ii) Popular databases (e.g., science news of daily newspapers) for mention of the primary journals in which the research was published; and (iii) Leading scientists in the specific field of interest were identified and consulted for possible suitable articles. Expect to screen dozens of articles before a few possibilities are examined in-depth for adaptation. Once a few possibilities are selected, the pros and cons of each are listed and at least three experienced (>10 years of teaching experience) in-service science teachers are consulted for choosing the best PSL article to be adapted as an APL.

As outlined in Chap. 2, APL is more closely related to PSL which is the main genre used for science communication among scientists. In addition to PSL, various other genres of scientific texts are used for communication among scientists, e.g. review articles, chapters in books (Goldman and Bisanz 2002). These text genres are not recommended as a basis for developing an APL article because they typically lack the essential features which characterize this genre (e.g., detailed description of the methodology used and presentation of the actual evidence that led the authors to reach certain conclusions). Consequently, the use of scientific texts other than PSL for adaptation will require an extensive adaptation which in turn will modify those texts entirely on the one hand, and will require the development of a rationale for a research paper on the other hand. Thus, a PSL article (or articles) is the most suitable source for developing an APL article, even though an APL article can be hypothetically developed using other scientific sources. The PSL article chosen for adaptation should preferably possess the characteristics of the PSL genre that are outlined in Table 2.1. Namely, it should be argumentative in nature, include evidence and reasons to support conclusions, have a canonical organizational structure, and present science as uncertain.

Choosing an appropriate PSL article for adaption to an APL includes the following eight nonsequential steps:

(a) *Adjusts to the future use of the adapted text*: When choosing an appropriate PSL for adaptation it is important to keep in mind the future use of the adapted text. Namely, whether it is adapted as learning and teaching materials that are planned to be used in the course of instruction of a specific curriculum in class, or whether it is adapted solely for research purposes.

(b) *Complements relevant curriculum content*: In cases where APL articles are adapted as teaching and learning materials, it is important to choose articles that focus on a topic which is included in the relevant curriculum in the schools or districts in which it will be implemented. For example, both developmental biology and biotechnology were listed as elective topics in the curriculum for 11th–12th grade biology majors in Israel, but no teaching and learning materials were available at the time the work on the APL-based materials for those topics was initiated.

(c) *Supplements the instructional sequence*: In cases where APL articles are adapted with the aim of integrating them into an existing curriculum, it is important to choose articles that focus on topics that appear in the curriculum or complement topics that already appear in the curriculum. For example, two APL articles were developed to supplement each of the three core topics of the high school biology curriculum in Israel, namely the living cell, systems in the human body and ecology (Israeli Ministry of Education 2011). The rationale for the integration of APL into existing curricula, which are not APL-based, is to enable the integration of authentic scientific texts into the teaching and learning of biological topics that are mainly taught and learned using textbooks. Moreover, it can be used as a means to update the curriculum with new scientific findings, since the APL articles can be easily replaced on an annual or bi-annual basis.

(d) *Establishes the credibility of the sources*: In cases where APL articles are adapted as teaching and learning materials of an entire topic, it is advisable to choose PSL articles published in leading journals with a relatively high impact factor, thus securing their relative importance to the field as well as the credibility of their sources. For example, from the curriculum in Biotechnology for 12th grade biology majors (Falk et al. 2003), we chose to adapt an article about the first success of stem cell gene therapy published in *Science* (Aiuti et al. 2002). In cases where the APL articles are planned to supplement an already existing topic in which the logic and reasoning are built using other learning materials (i.e., a textbook), it is possible to choose PSL articles which are less central to the specific field.

(e) *Matches materials to students' prior knowledge*: Students' prior knowledge is an important consideration throughout the adaptation process of an APL article. It is especially important to examine carefully the Methods used in the PSL and consider whether it will be possible to explain the principle of each method (or of a few of them) to the target audience. Similarly, it is important to examine carefully the various illustrations used in the Results section of the PSL, and whether it will be possible to include them, or modifications in the APL, to heighten the chances of student understanding. For example, when considering suitable PSL articles for adaptation to a curriculum in Biotechnology for 12th grade biology majors (Falk et al. 2003), we eliminated PSL articles in which the DNA or RNA electrophoresis technique was used, because we knew that most 12th grade biology majors are not familiar with this technique and that it might be too abstract to be learned solely from the text. In contrast, techniques such as polymerase chain reaction (PCR) or enzyme linked immunosorbent assay (ELISA), which are also abstract but more familiar to the students from their prior studies, were found suitable for inclusion in the APL articles that were included in the curriculum. Such a criterion is fundamentally important when choosing the PSL for adaptation.

(f) *Provides a clear and logical research approach*: It is preferable that the chosen articles for an APL-based curriculum describe either a one-step experiment or a few closely-related steps. In this way, as also reported by Bandoni Muench (2000), the logic of the research plan is easier to follow.

(g) *Provides visualization to complement the results*: Articles which describe results in simple, direct prose combined with explicit and complementary visualizations are preferred (Phillips et al. 2010). For example, in the developmental biology curriculum one research article demonstrates the control of finger pattern during limb development by manipulating finger duplication, which is easily illustrated in the figures of the original publication (Riddle et al. 1993).

(h) *Provides variety of research subjects, experimental organisms, and research approaches*: In cases where APL articles are adapted as teaching and learning materials of an entire topic they serve as the main source of information for learning that topic. Accordingly, when choosing the PSL articles for the curriculum in developmental biology (Yarden and Brill 1999), the articles were chosen so that altogether they would present a variety of the following: (i) *research subjects*: each article demonstrates research in one of four topics in developmental biology: differentiation (Hasty et al. 1993), embryonic induction (Riddle et al. 1993), genetic control of development (Driever and Nusslein-Volhard 1988; Nusslein-Volhard et al. 1987), and cell migration (Le Douarin and Teiller 1974; Le Douarin et al. 1975). For each topic, the classical research in which the main breakthroughs were made was chosen; (ii) *experimental organisms*: each article demonstrates research in which a different organism was used as an experimental system (chick, mouse or fly); and (iii) *research approaches*: each research uses a different approach to answering the research question (genetic, molecular, classical embryology or their combination).

(i) *Anticipates students' motivation to and interest in reading the text*: Students' interest and motivation to read the adapted text are of utmost importance to consider when choosing an appropriate PSL for adaptation. Scientific topics that are relevant to the students' lives are known to elicit their interest and motivation to learn. For example, at the time we chose the Mourez et al.'s (2001) article for adaptation for research carried out by Baram-Tsabari and Yarden (2005), envelopes with a white powder suspected as anthrax spores were reported on the daily news to have been sent to various locations around the world. We anticipated that a text telling about a possible scientific solution to this current and newsworthy biological weapon would be of interest to the students and in turn motivate them to read the texts on the topic that were presented to them. Different considerations led us when choosing suitable PSL articles for adaptation for a curriculum in Developmental Biology for 12th grade biology majors (Yarden and Brill 1999). In this case, we looked for articles that report on studies considered breakthroughs in the field of developmental biology. We aimed to choose articles with high-level claims deemed to be more important, than those with low-level claims and hence considered to be more trivial (Swales 2001, p. 117), and we anticipated the former will motivate students to read. Towards this end for the Developmental Biology Curriculum, one of the articles we chose was written by Christiane Nüsslein-Volhard, who won the Nobel Prize in Physiology and Medicine together with Edward B. Lewis and Eric F. Wieschaus in 1995, for their breakthrough research on the genetic control of embryonic development (Driever and Nusslein-Volhard 1988; Nusslein-Volhard et al. 1987).

Adapting Research Articles

In the previous chapters, we have mentioned that PSL is highly professional and complex in nature, written by experts, and difficult to read for novices. The reasons for these difficulties lie mainly in the unfamiliar professional and technical language, the density of details given in contrast to the limited amount of knowledge culled, and the fact that students (at both secondary or college levels) are unacquainted with research methods. In addition, scientists often demonstrate their results using phenomena outside of the everyday experience of high school students (for example, bands on a gel are a common representation in research articles in biology). As a result, students as novice science readers have to go through several intermediate steps to understand the conclusions from the data (Bandoni Muench 2000).

To overcome the gap between the highly professional and complex nature of the PSL articles and the cognitive level of high-school students, the original PSL article chosen for adaptation undergoes rewriting. Nevertheless, the repeating basic structure (abstract, introduction, methods, results and discussion) typical of most research articles, is retained. Moreover, the function of each section of the PSL is retained. As outlined by Hill et al. (1982) and represented in Fig. 5.1, the overall organization of the article is from the general to the particular and back to the general. Specifically, the Introduction section of PSL usually provides a transition from the larger academic field to the particular experiment and states the purpose of the research, the Methods and the Results sections specify the particular procedures and the nature of the findings, and the Discussion usually guides the reader from the particular experiment back to the larger academic area (Fig. 5.1). This overall functional structure is maintained in the adapted articles, or added if missing in the original PSL article.

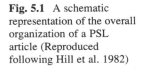

Fig. 5.1 A schematic representation of the overall organization of a PSL article (Reproduced following Hill et al. 1982)

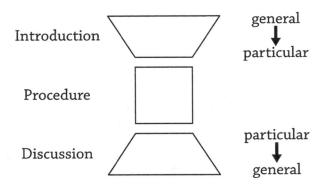

We detail next the adaptations that are typically made to a PSL article(s). The adaptations are listed according to the various parts of a typical article, and include specific examples.

(a) *Title*: Since the titles of PSL articles often include professional terms that are unfamiliar to the high school students who form the target population of APL articles, the titles are slightly modified during the adaptation process. The modification usually eliminates the technical and professional terms (see underlined text in Table 5.1), yet retains the essence of the original titles (Table 5.1). The first two examples of titles in Table 5.1 transmit to the reader the main findings of the research described in each of the scientific articles, namely reporting about a scientific achievement, while the third example simply reports about the development of a product, namely about a biotechnological achievement. These messages are retained in the titles of the corresponding APL articles (Table 5.1).

(b) *Abstract*: The abstract of an APL article is re-written completely at the last stage of the adaptation process. Since the original PSL article is modified extensively during the adaptation, the original abstract usually does not represent the essence of the re-written APL article. While writing the abstract of the APL article the internal invisible structure is retained: one to two sentences representing each of the various sections of the article, Introduction (including objective/research question), Methods, Results, and Discussion, are included in the abstract.

(c) *Introduction*: The introduction section of a PSL article usually provides the rationale for the article, and moves from a general discussion of the topic to the particular question, issue or hypothesis. A secondary purpose of the introduction section is to attract the readers' interest in the topic (Swales and Feak 2012). Those features are kept in the introduction of the corresponding APL article. However, during the adaptation process the introduction section of the APL is re-written to fit the knowledge level of the target population. Initially, the prior knowledge that is required for comprehending the article is mapped. Toward this end, a list of terms and processes that are essential for understanding the specific PSL article is prepared. Then, the envisioned prior knowledge of the target population with regards to the specific topic of the PSL is mapped, using the common standards in the specific region or country

Table 5.1 Examples of titles of PSL and the corresponding APL articles

PSL	APL
Muscle deficiency and neonatal death in mice with a targeted mutation in the myogenin gene (Hasty et al. 1993)	Absence of skeletal muscles in mice embryos carrying a mutation in the myogenin gene (Yarden and Brill 1999)
Overexpression of the Bt cry2Aa2 operon in chloroplasts leads to formation of insecticidal crystals (De Cosa et al. 2001)	Expression of the Bt bacterium toxin in chloroplasts of tobacco plants imparts resistance to insects (Falk et al. 2003)
Designing a polyvalent inhibitor of anthrax toxin (Mourez et al. 2001)	Developing an inhibitor of anthrax toxin (see Part III, Chap. 8)

Terms which were identified as potentially unfamiliar to high-school students are underlined

(for example the national syllabus for high school biology in Israel, Israeli Ministry of Education 2011, was examined in order to identify whether the required terms and processes are present). A comparison between the knowledge that is required to comprehend the article and the envisioned prior knowledge of the target population is used as a basis for planning the content that will be explained in the introduction section, and is omitted from PSL. Accordingly, the introduction section of the PSL is modified to give the novice reader basic background information which was either omitted from, or stated in the original PSL article. If an experiment was mentioned in the original PSL article with reference to another article, it is explained and sometimes even illustrated with a figure. The example provided in Table 5.2 shows how a single sentence from a PSL was adapted to fit the knowledge level of the readers of the corresponding APL. Such an extensive adaptation is not carried out for all of the sentences that appear in the introduction section of the PSL article, only those that are essential for understanding the message of the specific article are explained in detail in the APL, and the others are omitted. In this case, the adaptation process provides an overview of several crucial experiments that sets the groundwork for the experiment described in the specific PSL, and attempts to provide possible controversy(ies) in the field that justify the specific study. As the readers of APL are usually not familiar with the literature as are the readers of PSL, and usually they do not have access to the references that are cited in PSL, the essential background knowledge is provided in the APL. An important point to keep in mind while writing the introduction section of the APL article is to keep the overall organization of the section, so a transition from the general to the particular can be easily identified (Fig. 5.1). In addition, the introduction should end with the objective of the study, and/or the research questions addressed in the article. The questions are often reformulated to fit the modifications made to the article and the knowledge level of the target population (Table 5.3).

(d) *Methods*: The methods section of a PSL article usually describes in varying degrees of detail, methodology, materials and/or subjects, and procedures (Swales and Feak 2012). In the APL article, only the main principle of a method is described, while details of amounts, solution compositions, etc. are omitted. The reason for the different emphases given in the methods section in

Table 5.2 An example of an adaptation (A to D) of one sentence from a PSL article

A. On the basis of their activities in tissue culture, it was anticipated that these proteins would play a crucial part in the differentiation and maintenance of skeletal muscle in mammalian embryos[1–4]

| B. In an experiment performed on **cell cultures**, each one of the myogenic genes has been inserted separately by genetic engineering methods into cells that were not muscle cells. As a result, these cells differentiated into muscle cells similarly to the differentiation process of muscle cells which normally occurs in the entire organism (Fig. 1a–d): first, they resembled embryonic muscle cells (Myoblasts[a], Fig. 1b), then, they expressed muscle specific genes, and fused to become multi-nuclei muscle cells (Myotubes, Fig. 1c), and finally, created adult muscle fiber (Myofiber, Fig. 1d), and even started contracting in culture | C. **Cell Culture**: A method which allows growing cells in controlled conditions outside the living organism. Cell cultures allow performing experiments that are not possible in the whole organism, in a controlled way |

D.

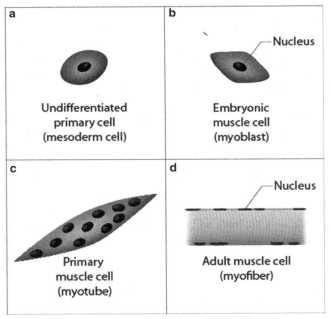

Fig. 1 Schematic illustration of the differentiation process of an adult muscle cell. The differentiation process starts from an undifferentiated primary cell (**a**) from the mesoderm that differentiates into embryonic muscle cell (myoblast); (**b**) The shape of the myoblast in culture resembles a star. Further on in the differentiation process, some myoblasts become confluent, lying alongside each other, and fuse to create a primary muscle cell (myotube); (**c**) A myotube is a multinucleate cell (syncytium). High levels of actin and myosin proteins are synthesized in the myotube, and the cell transforms into muscle fiber (myofiber); (**d**) Along the myofiber, actin and myosin fibers, which are responsible for fiber contraction, are observed

(continued)

Table 5.2 (continued)

A. A sentence taken from the Introduction to Hasty et al. (1993). The numbers 1–4 at the end of the sentence refer the readers to the list of references at the end of the PSL. Such referrals are usually omitted from APLs

B. A paragraph from the Introduction to the corresponding APL article (Yarden and Brill 1999). In this text the term "cell cultures" appears in bold, referring the readers to a definition of the term that usually appears in the margins of the same page. There is also a referral to a footnote, clarifying the meaning of the term "myoblast", which appears at the bottom of the same page
[a]Myoblast: *myo* Greek for muscle, *blast* bud

C. A definition of term that appears in the margins of the text

D. A graphic illustration along with an explanation that accompanies the paragraph shown in B

Table 5.3 An example of the objective for a study as it appears in the PSL article and in the corresponding APL article

PSL	APL
To test *myogenin*'s role in embryonic development directly, we generated mice homozygous for a targeted mutation in the *myogenin* gene (Hasty et al. 1993)	In this study, we investigated the influence of the absence of *myogenin* on mice embryonic development (Yarden and Brill 1999)

Terms identified as potentially unfamiliar to high-school students are underlined

APL is that in a PSL article all the technical details of the procedure are provided, in order to enable other scientists to repeat the experiment(s) should they wish to do so. Those details are likely meaningless to readers of the APL and will not help them comprehend the article. However, the readers of an APL do need to understand the main principles of the method(s) used, as those are essential to comprehend the article. Those principles are usually omitted from a PSL article, because they are common knowledge among experts in the field, the target population of a PSL article. The methods section of an APL article is usually divided into sections and describes in general the subjects as well as the principles of the methods used, which are often accompanied with an illustration that does not appear in the PSL. For an example of such an illustration, see Fig. 5.2.

(e) *Results*: The results section of a PSL article usually describes the findings, along with variable amounts of commentary (Swales and Feak 2012). In the APL article, results which are sidelights to the main research question are omitted. In cases where the results have diminished importance to the students and require a lot of unfamiliar prior knowledge to understand—they are also omitted. For example, the first figure in the PSL article of Hasty et al. (1993, p. 502) demonstrates the successful insertion of a foreign piece of DNA into the *myogenin* locus within the mouse genome. The figure includes a detailed description of the molecular manipulations as well analyses at the DNA, RNA, and protein levels, thus providing a proof that indeed knockout mice were created. The figure was completely omitted from the corresponding APL (Yarden and Brill 1999) because it is too complex for high-school students to understand, and it does not add significant data to the main research plan or to the main results. The main figures are usually kept, with slight modifications and omissions. For example, in contrast to the PSL (Hasty et al. 1993), only normal

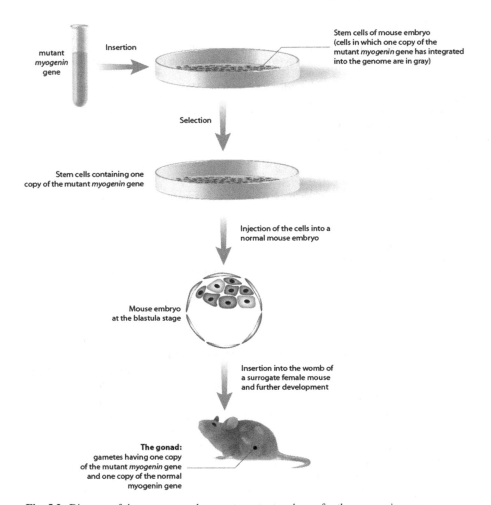

Fig. 5.2 Diagram of the process used to create mutant embryos for the *myogenin* gene

and mutant mice embryos are shown in the corresponding APL and the heterozygous embryos (those carrying one normal copy and one mutated copy of the
gene) are omitted. The reason for that omission is that heterozygous embryos
developed similarly to normal embryos, while homozygous embryos (those
carrying two copies of the mutant *myogenin* gene) were unable to move and
died shortly after birth. The contrast between the homozygous mutant embryos
and the normal embryos is dramatic, while omitting data related to the heterozygous embryos allow simplifying the results presented in the APL. In addition,
arrows were sometimes added to a photograph in the APL to help the students
recognize specific parts of the embryo or the tissues shown (Fig. 5.3).

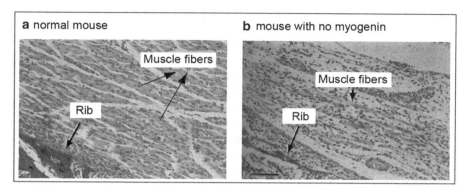

Fig. 5.3 A figure taken from the results section of the APL article: Absence of skeletal muscle in mouse embryos carrying a mutation in the *myogenin* gene (Yarden and Brill 1999). This figure includes two of ten images that appear in Figure 3 in Hasty et al. (1993, p. 504)

(f) *Discussion*: The Discussion section of a PSL gives meaning to and interprets the results in a variety of ways, at least some of which refer to statements made in the Introduction (Swales and Feak 2012). As with the introduction to an APL article, details are added to parts of the discussion of an APL article in order to match the prior knowledge of the target population and allow readers to understand it more easily. For example, if the researcher based a future experiment on a certain assumption, this assumption is explained at the students' level of understanding. Only the parts of the discussion section of the PSL that are relevant to the results shown in the APL are included, with the required adaptation, in the adapted article. While writing the discussion section of the APL it is built it in such a way that the internal organizational structure will clearly go from the particular to the general. In addition, open questions for future research as well as the study limitations are included in the discussion. In cases where those features were not included in the discussion of the PSL article, they are written for the APL article. As pointed out in Chaps. 2 and 3, one of the unique features of PSL is its argumentative structure. Accordingly, the discussion section of an APL article should include aspects of argumentation that will enhance the readers' understanding of the uncertainty of science as expressed in scientific texts. Therefore, it is advisable to include in the discussion section of an APL article other works that show contrasting evidence and claims to those that are put forward in the APL, and to discuss their meaning. In addition, statements about possible controversies in the field that were outlined in the Introduction section are discussed in this section in light of the findings of the current PSL study.

(g) *Other additions*: When adapting a PSL article, as described in the preceding sections a to f, several other additions are often included in the APL: (i) a paragraph is sometimes added at the end of each APL article with an explanation of the contribution, relevance and importance of the work described in the article to the understanding of scientific processes. This paragraph is usually designed to resemble the description of research articles in the "news and views" sections in professional scientific journals (e.g., *Nature*). Such descriptions are usually written by scientists who did not participate in the research, to give a global view of the research and its importance. This addition may help the students to understand the research that they are reading about in the correct context, which is very familiar to the experts in the field, but is completely new to the novice readers; (ii) definitions and explanations of terms are added in the text margins right next to their first mention in the text (as shown for example in Table 5.2, part C); (iii) questions to think about are sometimes added in the text margins to refine a certain point being made in the text and to challenge students to think about the research from multiple perspectives. As a result of these additions and modifications to the original PSL articles, each APL article both focuses on a single issue and provides the extended theoretical background needed for high-school biology students to understand the research to be read and discussed.

A Summary of the Suggested Sequence of Steps for Creating an APL

Once a suitable PSL article is chosen as a candidate for adaptation (as described above), the PSL article(s) should be read carefully. It is advisable that the expert (s) who will adapt the article perfectly understands the text and is familiar with its cited literature. Based on our experience, sometimes the experts are part of the scientific team who conducted the study and wrote the PSL (Norris et al. 2012), sometimes they are experts in the specific scientific discipline (Yarden and Brill 1999), and sometimes they are experts in science education research with a strong scientific background (Zer-Kavod and Yarden 2013). At this stage, it is important to identify the relevant background knowledge needed by the target population in light of the knowledge that is required for comprehending the article and keep in mind the possible concepts and processes that may require further elaboration beyond the information provided in the text of the chosen PSL.

The subsequent recommended step is to choose, from the Results section of the PSL, the results to be included in the adapted article and to make sure that altogether the chosen results form a coherent "story" of the specific scientific investigation. At this stage, it is important to make sure that the elimination of parts of the results that are included in the original PSL article will not impair readers' understanding of the scientific investigation. Accordingly, it is possible to choose a relatively long PSL article with several important results, and at this stage to make the decision on which parts of the results to include in the adapted article and which parts to eliminate. The next recommended step is to choose the representations of the results to be included in the APL. Once all abovementioned preparations are made the writing of the APL can begin.

Initially, it is recommended that the Results section of the APL be re-written to align with the chosen representations. Subsequently, the Methods section can be re-written and the appropriate representations prepared for inclusion in the Methods section. There will be several back and forth rewrites between the Results section and

Step number	Action taken
I.	Choose a PSL article
II.	Identify background knowledge
III.	Choose results
IV.	Choose data representations
V.	Re-write Results section
VI.	Re-write Methods section
VII.	Re-write Introduction section
VIII.	Re-write Discussion section
IX.	Re-write Title
X.	Re-write Abstract section
XI.	Obtain feedback from teachers
XII.	Obtain feedback from students

Table 5.4 A summary of the suggested sequential steps in creating an APL article

Note that steps V to X are iterative

the Methods section, as well as alignments of those sections with the level of relevant background knowledge of the target population, and then the writing of the Discussion and the Introduction sections can begin. All sections are then aligned, a title is written and finally an abstract. It is recommended that when the sections of an APL article are completed, to seek the advice of experienced teachers and science educators in order to improve the text. Finally, implementing the reading of new APL articles in schools and examining its outcomes are the most important means to obtain additional comments about the text for continued improvement and fit for the target population (Table 5.4).

Instructional Approaches Developed for the Use of APL in Schools

We outline several instructional approaches we developed for the use of APL in schools. These approaches are based on numerous conversations with teachers who were at first reluctant to use APL with their classes, but over time they acquired confidence and looked for additional such texts for instruction. Some of the outcomes of these conversations in schools are outlined in Chap. 7. Next, we first discuss the main approach that was developed and used in schools, the conversational approach, and subsequently we discuss approaches developed for instructing specific sections of the APL.

A Conversational Approach to Learning Through Research Articles

Reading is a constructive process that is iterative, interactive, and principled and shares the features of inquiry (Norris and Phillips 2008). We wondered whether this dynamic nature of reading could be used with a whole class of readers? and whether we are able to create a community of readers in class? In order to do that, an approach for learning through APL in the classroom was constructed. According to this approach students conduct a "conversation" with the article, rather than read it individually, in order to obtain as much knowledge as possible (Yarden et al. 2001).

According to this approach, students read the article together in the classroom one section at a time. After each section, the students raise questions about the part they read, and the teacher writes the questions on the board, or an interactive whiteboard (IWB), or on a transparency. The questions can be either clarifying questions or research questions. In order to answer at least some of the questions, students propose hypotheses or predict the outcomes of suggested experiments. By reading the next section together, the students can obtain answers to some of their questions, and can verify their predictions in the subsequent class discussion. In this manner, students proceed step by step with the reading while resolving themselves, as much as possible, their reading comprehension problems and the questions they raised.

To create an open atmosphere, students are asked to sit in a circle, with the teacher sitting among them. This way of sitting creates a feeling that the reading is shared, and everyone (and also every comment) is equally valued. During the first discussion phase, the teacher assumes the task of leader, and as the discussion progresses, the teacher's role evolves into moderator, rather than instructor. This approach to reading a research article was introduced to 11th grade high-school biology-major students during implementation of the APL-based developmental biology curriculum (Yarden et al. 2001) and the biotechnology curriculum (Falk and Yarden 2011). The outcomes of these experiences are described in Chap. 7. The conversational approach includes the following iterative stages:

1. A student reads the Title, a short section of the article or several paragraphs of a longer section (hereafter referred to as Section "A Step-By-Step Description of How to Develop APL") aloud. The teacher then asks: *What can we learn from the title and what questions does it raise?*
2. Students raise questions about Section "A Step-By-Step Description of How to Develop APL". All the questions, without discrimination on level of sophistication or the correctness of their assumptions, are recorded on the board.
3. Students make predictions about the way in which some of their questions will be further investigated by the scientists and about the sections of the article that will include the answers to their questions. This second type of prediction is preceded by an exploration of the structure of the scientific article and the function of its different sections. At this stage, the teacher avoids supplying answers to students' questions or making comments on the correctness of their predictions. For example, the teacher can ask: *Please, read the Title [again]. Do we expect the article to deal with this topic? Why, and Where [in what article section] will we find information about this question, if at all?*
4. By reading the next section or paragraphs, students may find answers to their previously formulated questions and compare the described research with their predictions. The teacher can then ask: *Did the Abstract answer some of your*

questions? Re-read and try to raise a question to which the answer is found in this paragraph.

5. Finally, students formulate new questions and the process (stages 1–5) is repeated.

It is important to note here that teaching via the conversational approach emphasized a few characteristics of the model:

(a) *Encouraging students to ask questions:* The approach is based on learning via questions that the students raise. However, it is well established that questions during a lesson are usually asked by the teacher, while students hardly raise any (Dillon 1988). Questions do not emerge spontaneously from students in the classroom; rather, they have to be encouraged (Dillon 1988).

The content of a question can indicate the level of thinking of the student who raised it. Usually the cognitive level of a question is determined by the type of answer it requires. A few ways to categorize questions have been previously suggested. Most researchers agree questions that require declarative knowledge as an answer, i.e. informative questions, indicate a low level of thinking (Dillon 1984; Shepardson and Pizzini 1991; Dori and Herscovitz 1999; Cuccio-Schirripa and Steiner 2000). When students ask questions during the lesson, they are usually informative ones (Dillon 1988).

In the conversational approach for learning through research articles, the first questions to be raised are usually about the article's title. At this stage of the reading, none of the students are familiar with the content of the article. Therefore, the questions that are raised are usually informative and about the meaning of terms in the title. Writing these questions as a list (on the blackboard, or on a transparency, or on an IWB) provides clear support for the legitimacy of the questions, as simple as they may be, showing the students that asking any type of question is important. It also later serves as a reference to show the students that they can answer these questions by themselves after further reading of the article. In this way, students are motivated to ask questions (and construct answers), and the teacher exhibits approval of question-asking (Dillon 1988). Undoubtedly, the way the students and teacher are seated also contributes to the open atmosphere and stimulates students to ask questions. During further reading, especially after reading the Introduction and the Methods' sections, informative questions become less and less frequent; nevertheless, the structure of reading-asking-reading and more-answering-asking is maintained.

(b) *Discussing basic terms:* Research articles present a good opportunity to clarify some of the basic terms that appear in the text. In the course of reading research articles, using the conversational approach, the students themselves raise questions about some of the basic terms, since they do not differentiate between new terms and "already learned and forgotten" ones. Therefore, these questions can be discussed and clarified in a supportive atmosphere.

(c) *Low-ability students:* We were amazed to find that the conversational approach also stimulates the supposedly low-ability students to ask questions and participate in the class discussion. Their participation is probably due to the atmosphere created by the circled seating, and the encouragement obtained by writing and immediately answering any question, no matter the type or by whom it was raised.

(d) *Scientific conversations between students:* Students find this forum for discussing a research article open enough to talk about issues among themselves. For instance, one student may raise a question that another student will answer. Disagreements may occur and students will debate an issue. In these situations, misconceptions may be revealed and dealt with by the teacher.

(e) *Predicting the main experiment:* One of the main characteristics of the conversational approach for reading a research article, in addition to asking questions, is the ability to predict the next section of the text. During reading and discussing the Introduction of the article, students are engaged in understanding the academic background of the research they are about to read. At the end of the introduction, after the scientist introduces the main research question, some students will suggest an experiment to answer that question. This response can happen almost automatically, without any prompting, since reading and discussing engage the students to suggest ways of solving the raised problem.

(f) *Different topics and levels of discussions:* Each part of the research article can serve as a basis for discussing different issues. For example, while reading the methods, students may be asked to recall other methods that could be used to the same end, and discuss the reasons why they were not chosen for the experiment they are reading about. At the stage of reading the results, students can try to analyze the data, and discuss their relevance to the research question. At the stage of reading the discussion section, students can relate the data to a model proposed by the scientist or try to raise new questions that could be answered by future experiments.

(g) *Identifying misconceptions:* The presentation of methods and results of authentic experiments in the format of a research article may be useful towards discussing students' experiences in the laboratory and the misconceptions that may ensue. For example, an interesting remark was made by one of the students while reading the methods of a particular research article, when discussing the reasons for publishing the methods so accurately, the issue of criticism was made (among other issues), and the teacher commented, "*It is important to see whether there is some kind of discrepancy between the method and the research question, namely that the method is "not good" for this specific question, and therefore the results are not valid.*" The student then asked: "*But how can one obtain results when the method is not good?*" This remark demonstrates a basic misconception. The student is relating the research in the article to her own experience in the school laboratory, where results not matching expectations may indicate to her that she did something "wrong" during the experiment ("the experiment didn't work"). This is an example of an excellent opportunity to discuss such misconceptions, which may emerge while discussing an article.

Because reading a research article is not a skill that students can learn by themselves, the teacher plays a very important role in teaching through the read-reread-conversational approach of the APL research articles. It is also important to encourage question-asking. If students are uncomfortable with asking questions, then the conversational approach cannot proceed. The teacher can also help the students with guiding questions he or she has formulated. These questions can either mimic the expected student questions or help students integrate the information obtained from the article with their relevant background knowledge. These questions can help the students while reading an article by themselves, and can be discussed later in the classroom in order to facilitate understanding. The teacher can use research articles to develop scientific literacy among the students. Issues that concern the nature of science, like the importance of describing the methods clearly and accurately; the way research is evaluated in the academic world; the importance of publishing the results of research; and the time needed to perform experiments to try and answer a single research question, can be raised and discussed. One of the main characteristics of research articles is their highly professional and technical language and structure. The teacher can emphasize the differences between research articles and popular articles, and help the students to recognize the importance of communicating scientific data clearly and accurately. For the student, the article can be an isolated piece of knowledge. The teacher should help the student make as many connections as possible between the knowledge acquired from the article and other subjects and issues that are learned in class. Possible connections could be the social, ethical and philosophical implications of research conducted in the specific field. Another characteristic feature of research articles is the density of details provided, in contrast to the seeming relatively small piece of knowledge obtained. The following citation, taken from a lesson in one of the classes that implemented the APL-based curriculum in developmental biology (Yarden and Brill 1999), demonstrates how a teacher helped her students grasp the effort needed to answer a single research question:

T: We have now finished reading and learning our first research article. Who can summarize the conclusion from this research?

S: That... the neural crest cells... that they can migrate according to the environment in which they migrate.

T: Good! You summarized the article in one sentence. You see, if we open a text book, this whole article would add this one sentence to the book. That's all. But you can see now how much work is needed to prove this result!

Opening the Teaching and Learning of APL

Lawson (1988) suggested that the opening stage of a learning cycle should promote students' exploration of a specific example by direct observations and raising questions. An important question is whether the pedagogical aims assigned to the opening stages of a scientific topic in class are also applicable to the opening of an APL article and if so, what models may be suitable for opening the teaching and learning of an APL article? What scientific reasoning processes do these teaching/learning models elicit at the students' level and what challenges do they present for the students and the teachers?

The instructional strategies for the opening sections of APL were designed while taking into account the following considerations: (a) students might be intimidated by the sudden exposure to an overwhelming number of novel terms and ideas; (b) content aspects might be given priority over aspects of inquiry and NOS understanding, due to the challenging text and the numerous questions raised by the students; (c) teachers, although agreeing to the importance of inquiry learning, might revert to teacher-centered approaches while attempting to facilitate students' challenged comprehension.

In addition to the *conversational model*, which was described in detail above, two additional approaches were developed: the *problem solving approach* and the *scientific literacy approach* (Falk and Yarden 2011):

I. The *problem-solving approach* of enacting the opening sections of an article involves presenting the students, before their first exposure to the article, with a problem similar or tangential to the one that the scientists are exploring and asking students to suggest suitable methods and experiments to solve it. During the ongoing collaborative process, students' answers are based on both their prior knowledge and inquiry skills and new, relevant information received from their teacher, according to the advancement of the solution. Therefore, this approach is to some extent similar to the teaching of "Invitations to inquiry" (Schwab 1962). Its main stages are:

1. The teacher presents a problem similar to the practical or research problem investigated in the article. If the problem has social implications, they are explored in order to expose the students to the diverse aspects that may affect the problem-solving. Students' prior knowledge should allow them to understand the problem and afford them the ability to offer compatible solutions.
2. Students suggest different strategies for solving the problem, taking into consideration as many parameters as possible.
3. The teacher guides the students by supplying new relevant information, posing questions or suggesting alternative ways of solving, and asks them to analyze the rationale and relative advantages of these solutions.
4. When a tentative solution is reached (it does not necessarily have to match the solution proposed by the article), the teacher directs the students to read the opening sections of the article.

   For example, a teacher in one of the classes implementing an article about biosensors from the APL-based curriculum in biotechnology started the instruction (stage 1) of the article by asking: "We have just established the Water Management Council. You are the managers and you have all the scientific help you need. Now, you suspect contamination of a drinking-water source. How do you proceed?" Subsequently, students suggested different means for solving the problem (stage 2), and the teacher guided the process (stage 3) by saying: "Let's try to find another way. Maybe it's possible to do it faster and at lower cost." At stage 4, the teacher directed the students to read the opening sections of the article and gradually focused on specific strategies and the methods that could be used. The teacher guided the students to suggest improvements to their suggested strategies, thereby reflecting a main tenet of biotechnology. He opposed some of the students' suggestions by pointing out the practical and experimental limitations (e.g. the polluting substance is unknown, therefore one cannot use a specific indicator) and wrote on the board a summary of some of the principles applied by the students to solve the presented problem.

At the beginning, students' comments were highly unfocused and based mainly on their practical experiences in the school laboratory or in everyday life. For example, one students said: "I would check the water." Another student said: "[One should] see the pH level, maybe bacteria are growing." Gradually, the students' comments became more specific, taking into consideration the possible limitations of their suggested solutions: "The question is if it [the reaction of animal biosensors] is the same in humans." In the last stage, although the students did not suggest the use of bacterial biosensors by themselves, they could support the rationale for their use, when suggested by the teacher. "They are smaller [the bacteria], and smaller quantities [of genotoxic materials] will affect them."

II. The *scientific literacy approach* focuses on a comparison between different genres of articles in terms of aspects of scientific communication—the need to publish the results of one's research, their different audiences and characteristics, the information provided by each genre and their different communication styles and suitability to the relevant audience. This approach is not as rigorously structured as the problem-solving approach and the conversational approach described above. The scientific literacy approach exposes the students to the same scientific topic presented in two or more articles belonging to different genres. Usually, the information provided is complementary: while the popular reports present ideas and concepts in a narrative genre and in a less technical language, thereby facilitating comprehension, the APL articles provide more detailed information that helps elucidate the inquiry aspects (Baram-Tsabari and Yarden 2005). The sequence of the exposure may vary. Some teachers use the popular reports before the teaching of the APL articles: students' questions on these reports are later answered by the information provided in the APL article and the comparison between the two genres is prompted after the students have accomplished the learning of both articles.

Benefits and Limitations of the Instructional Approaches

Since the three presented approaches promote different desirable science-learning skills and processes and are not mutually exclusive, we do not suggest a preference for one over the others. The conversational and problem-solving-based approaches emphasize learning by inquiry through formulating inquiry questions, designing methods and planning experiments, the scientific literacy approach emphasizes a better understanding of the NOS. The conversational approach relies in part on the students being responsible for knowledge acquisition, and the other two approaches provide more opportunities for teachers to channel the class discourse toward aspects they consider important. Aspects concerning scientific communication can be discussed in the context of both the conversational and scientific literacy approaches, the problem-based approach, however, is oriented more towards discussing discipline-specific strategies and metastrategies. Both the conversational and problem-solving approaches take into account that even though students may suggest incorrect answers or pose questions based on erroneous assumptions, the teacher does not need to immediately intervene in order to correct them, but can rely on the text to do so. The scientific literacy approach presumes students' prior knowledge of the characteristics of another genre, based on the initial analysis of another text.

Nonetheless, each approach has limitations: (a) the conversational approach has been reported as time-consuming and sometimes tedious because of its iterative stages; (b) the problem-solving approach may enhance the teachers' role as knowledge provider and regulator, instead of delegating this role to the authors of the article; and (c) the scientific literacy approach may be problematic due to the difficulty involved in finding texts that belong to different genres but refer to the same topic. Indeed, most teachers whom have enacted the APL-based curriculum have reported the use of a variety of approaches, sometimes in the context of the opening sections of the same article, in order to maximize their different benefits and minimize their respective limitations.

We consider all three approaches to contain the pedagogical components included in the initiation stages of different learning cycles. In the conversational approach, posing questions on the article sections that have been read enables exploration of the text, the research described and students' prior knowledge. In the problem-solving approach, the exploration of the discussed problem and the different aspects of its solution that can be identified in the article provide the same benefits. For the scientific literacy approach, reading one of the genres first provides an immersion stage to be further used as a comparison with the additional genres. This parallel can be drawn even though most learning cycles have been designed for problem-solving and hands-on inquiry, and not for science-text-based learning as in our case. An exception is the application of the learning cycle to text-based learning by Musheno and Lawson (1999) who showed that students achieve better concept comprehension by reading a text that presents examples before introducing new terms, the latter is a typical characteristic of scientific articles.

Ways to Support Teachers in Using APL

Learning through APL requires novel and challenging modes of teaching, as science teachers' use of scientific articles in class is almost exclusively limited to secondary literature (McClune and Jarman 2012). The novelty of the teaching strategies required, stems from the fact that other than coping with new content knowledge, teachers are concomitantly faced with the promotion of skills which are associated with learning through research articles. To assist the teachers that implement teaching using APL and to convey our perspective on the pedagogical content knowledge (PCK) that we considered adequate for the teaching process (Shkedi 1998; Shulman 1986), we carried out workshops for teachers and developed teachers' guides for the APL-based curricula. The features of the workshops and the unique characteristics of one of the teachers' guides are described next in light of the ways they support teachers in using APL in class.

Workshops for Teachers Implementing the APL-Based Curricula

The workshops carried out during the implementation of the APL-based curricula were designed to support the teachers during the use of the materials in their classrooms. Workshops for each curriculum (developmental biology or biotechnology, Falk et al. 2003; Yarden and Brill 1999) were conducted separately. The participating teachers taught students the curriculum in 11th and 12th grades, who studied towards the matriculation exam in biology in urban high schools in Israel. The syllabus for the biology studies in Israel includes, in addition to basic topics, advanced topics (including the two abovementioned APL-based topics) designed for 30 h of teaching. Each workshop included the following components: (i) The conversational approach for teaching and learning through APL, as well as additional teaching approaches (see above), were presented to the teachers during developers'-teachers' meetings; (ii) A collection of guiding questions and activities for the introduction of the curriculum and for the APL articles as well as their answers was developed and discussed with the participating teachers; (iii) Teaching sessions, in different schools were video-taped. The video tapes were used during the workshops in order to follow up on the teaching process, identify students' difficulties, and illustrate the teaching process through APL to teachers with no previous experience in teaching through this genre. At a later stage, the episodes were edited and included in the curriculum guide (see below); (iv) Some of the difficulties met by students who learnt through APL were mapped and remedial measures were discussed with the participating teachers in order to enable them to overcome the problems that had surfaced; (v) During the workshops all the participating teachers were requested to implement the APL-based curriculum in their classrooms and discussions about their experiences were carried out.

A Teachers' Guide for One of the APL-Based Curricula

The teachers' guide for the APL-based curriculum "The secrets of embryonic development: Study through research" (Yarden and Brill 1999) is described here as an example. The main characteristics of the guide are the inclusion of a pool of questions and activities that can be edited by the teachers and a collection of video-taped authentic teaching episodes. Both are aimed at providing the teachers with environment-effective advice while enhancing their feeling of autonomy. The teachers' guide is aimed to provide a scaffold for the acquisition of the pedagogical and content tools provided and to support the teachers in the appropriation of these tools. Moreover, it is designed to support the teachers to devise their own tools oriented toward the same pedagogical aims, during the process of curriculum implementation while using the teaching models provided in the guide.

The guide was designed on CD-ROM as it enables presentation of information to the teachers: (i) authentic teaching episodes; (ii) visual models of molecular topics; (iii) a gallery of all the pictures in the students' book in a format that allows their use for presentations in the classroom; (iv) a web-quest assignment in bioethics; and (v) a pool of questions to and activities for the introductory section of the curriculum and to the research articles. In addition, the guide includes the following textual information: (i) an introductory section entitled: "Why and how to teach through research articles". This section presents our best-informed position about the need to expose the students to research articles and our recommendations about possible approaches to teach through APL; (ii) remedial measures for the difficulties encountered by the students; (iii) activities aimed at enhancing the understanding of different forms of scientific communication; (iv) a list of the main biological principles that appear in the articles; and (v) assessment questions from matriculation examinations.

The videotaped teaching episodes, lasting from 3 to 10 min were subtitled and accompanied by: (i) a description of the background of the recorded episode: school, class grade, previous articles studied by the students, the aim of the study session and the teaching sequence of the session including the recorded episode; (ii) an annotated description of the episode: stage-by-stage description of the events occurring during the episode (e.g. "The teacher waits for a long time before the first student answers"); (iii) pedagogical comments on some of the main didactic and cognitive processes which occurred in the episode. The teachers are explicitly told that these comments solely reflect the authors' views and they are prompted to elaborate upon their own interpretation of the events whenever they feel it necessary; and (iv) open questions, asking the teachers to compare the interventions in two or more episodes, or to elaborate on alternative teaching strategies. The episodes can be watched without viewing the accompanying texts or concomitantly with the scrolling down of one of the texts. They were designed in such a way that watching the teaching episodes and analyzing them may lead the teachers toward realistic expectations which may in turn enhance their self-confidence needed for successfully coping with the novel teaching environment of using APL. In addition, it may provide the teachers with context-sensitive modeling stemming from a "virtual" apprenticeship of the novice teachers with the "role models" that may lead them to critical analysis which can include an inherent comparison between the novice teacher's personal teaching style and the "role model" teacher performing in the episode.

A copious number of questions and activities were developed for each section of the introductory unit and the APL articles. The integration of a variety of questions into the teaching process through APL is aimed to facilitate the creation of a dialogue between the student and the content of the article. The logically structured order of the article sections sometimes conveys, to the novice reader, a false feeling of comprehension. The questions on the article are aimed to unravel this false feeling and to encourage students to look for a deeper understanding and for new connections between the ideas presented in the article and the students' knowledge (Brill et al. 2004). The teachers were encouraged to use only those questions and activities best suited for their aims and their students' cognitive level and to modify them using the standard computerized tools. Enabling the teachers to choose from a collection of questions and to modify them according to their needs was aimed to enhance the appropriation of the questions and their usage.

We categorized the questions provided according to the cognitive skills required from the student and the content field sampled by the question, as shown in Table 5.5. We found this categorization to be suitable for addressing the main aims of teaching through APL (Yarden et al. 2001): Acquaintance of the students with the nature of scientific research, understanding of the rationale behind the research plan and methods and critically asserting the goals and conclusions of the scientific research. Some of the activities are in accordance with the stages of the conversational model—formulating questions, making predictions and finding answers in the subsequent sections of the article. Several questions and activities were added as remedial measures after mapping students' difficulties in studying through APL.

Our investigation of the process of teaching using APL with the aid of the teachers' guide showed that the videotaped class episodes that are included in the guide facilitate the simulation of teaching situations that occur while studying through APL. In addition, our analyses of the teachers' comments on the pool of questions and activities provided clear evidence that we afforded them with an easily adaptable teaching tool and conveyed our PCK perspective. Details about those investigations are provided in Chap. 7. It thus seems that if APL is to be considered a constructive and important component of science education, a fundamental necessity is that students be able to read the adapted articles, which is the subject of the next chapter, Chap. 6.

Table 5.5 Sample questions from the teachers' guide

Category	Sample question or activity from the teachers' guide
Knowledge organization	Write down the differences between maternal and zygotic genes that determine the embryonic development, considering: The cells in which these genes are transcribed, the transcription time since the fertilization and the stage in which the protein products of these genes are active in the cell. (for the article "Genetic regulation of the developing Drosophila embryo head")
Inquiry skills	Design an experiment in order to investigate the hypothesis that during two and a half hours after fertilization, the embryo's genes are not transcribed. (for the article "Genetic regulation of the developing Drosophila embryo head")
Critical assessment of the article conclusions	"From this evidence it is possible to conclude that *myogenin* is not essential when cells begin to differentiate to muscle cells, but... without *myogenin* expression the differentiation will not occur." Write down evidence for each part of this statement from the Discussion section. (for the article, "Lack of skeletal muscles in new born mice bearing a mutated *myogenin* gene")
Application of the main ideas of the article in other contexts	The biotechnology company "Moneygen" reported a sensational success: The production of *myogenin* containing pills for athletes interested to improve their performances without sweating. The competing company "Musclegen" is also advertising a natural product intended for athletes: Their product is a natural plant extract that was shown to increase the *myogenin* production in mice embryos. As a gym fan, would you use these products? Justify for each of the applications. (for the article "Lack of skeletal muscles in new born mice bearing a mutated *myogenin* gene")
Understanding the methods rationale	Which of the following methods could be used in order to establish when and where the myogenic genes are expressed during embryonic development?
	a. extraction and analysis of DNA from different embryo tissues
	b. extraction and analysis of m-RNA from different embryo tissues
	c. in-situ hybridization of embryo tissues with c-DNA of a myogenic gene
	d. protein extraction and analysis
	e. in-situ reaction of embryo tissue with antibodies against myogenic proteins
	f. detection of muscle cells in different tissues. (for the article "Lack of skeletal muscles in new born mice bearing a mutated *myogenin* gene").
	Draw the molecular complex that is formed in embryos when using the c-RNA detection method. (for the article "Genetic regulation of the developing Drosophila embryo head")

(continued)

Table 5.5 (continued)

Category	Sample question or activity from the teachers' guide
Highlight the main developmental biology ideas of the article	Three groups of genes influence the embryonic development according to the following hierarchy:
	genes that regulate the expression of the *bicoid* gene
	the *bicoid* gene that encodes for the morphogene of the head formation
	genes involved in the head tissue formation and controlled by the product of *bicoid* gene
	Which group would you expect to be expressed earlier during the development?
	If a mutation occurs in a gene belonging to one of the three groups, in which group is it expected to have a more critical influence? (for the same article)

References

Aiuti, A., Slavin, S., Aker, M., Ficara, F., Deola, S., Mortellaro, A., Morecki, S., Andolfi, G., Tabucchi, A., Carlucci, F., Marinello, E., Cattaneo, F., Vai, S., Servida, P., Miniero, R., Roncarolo, M. G., & Bordignon, C. (2002). Correction of ADA-SCID by stem cell gene therapy combined with nonmyeloablative conditioning. *Science, 296*(5577), 2410–2413.

Bandoni Muench, S. (2000). Choosing primary literature in biology to achieve specific educational goals. *Journal of College Science Teaching, 29*, 255–260.

Baram-Tsabari, A., & Yarden, A. (2005). Text genre as a factor in the formation of scientific literacy. *Journal of Research in Science Teaching, 42*(4), 403–428.

Brill, G., Falk, H., & Yarden, A. (2004). The learning processes of two high-school biology students when reading primary literature. *International Journal of Science Education, 26*(4), 497–512.

Cuccio-Schirripa, S., & Steiner, H. E. (2000). Enhancement and analysis of science question level for middle school students. *Journal of Research in Science Teaching, 37*(2), 210–224.

De Cosa, B., Moar, W., Lee, S. B., Miller, M., & Daniell, H. (2001). Overexpression of Bt cry2Aa2 operon in chloroplasts leads to formation of insecticidal crystals. *Nature Biotechnology, 19*, 71–74.

Dillon, J. T. (1984). The classification of research questions. *Review of Educational Research, 54*(3), 327–361.

Dillon, J. T. (1988). Questioning in science. In M. Meyer (Ed.), *Questions and questioning* (pp. 68–80). Berlin: Walter de Gruyter.

Dori, Y. J., & Herscovitz, O. (1999). Question-posing capability as an alternative evaluation method: Analysis of an environmental case study. *Journal of Research in Science Teaching, 36*(4), 411–430.

Driever, W., & Nusslein-Volhard, C. (1988). A gradient of bicoid protein in Drosophila embryos. *Cell, 54*, 95–104.

Falk, H., & Yarden, A. (2011). Stepping into the unknown: Three models for the teaching and learning of the opening sections of scientific articles. *Journal of Biological Education, 45*(2), 77–82.

Falk, H., Piontkevitz, Y., Brill, G., Baram, A., & Yarden, A. (2003). *Gene tamers: Studying biotechnology through research* (In Hebrew and Arabic, 1st ed.). Rehovot: The Amos de-Shalit Center for Science Teaching.

Goldman, S. R., & Bisanz, G. L. (2002). Toward a functional analysis of scientific genres: Implications for understanding and learning processes. In J. Otero, J. A. Leon, & A. C. Graesser (Eds.), *The psychology of text comprehension*. Mahwah: Lawrence Erlbaum Associates Publication.

Hasty, P., Bradley, A., Morris, J. H., Edmondson, D. G., Venuti, J. M., Olson, E. N., & Klein, W. H. (1993). Muscle deficiency and neonatal death in mice with a targeted mutation in the *myogenin* gene. *Nature, 364*, 501–506.

Hill, S. S., Soppelsa, B. F., & West, G. K. (1982). Teaching ESL students to read and write experimental-research papers. *TESOL Quarterly, 16*(3), 333–347.

Israeli Ministry of Education. (2011). *Syllabus of biological studies* (10th–12th grade). Jerusalem: State of Israel Ministry of Education Curriculum Center (In Hebrew) http://cms.education.gov. il/EducationCMS/Units/Mazkirut_Pedagogit/Biology/TochnitLimudim/

Lawson, A. E. (1988). A better way to teach biology. *The American Biology Teacher, 50*, 266–278.

Le Douarin, N. M., & Teiller, M. A. M. (1974). Experimental analysis of the migration and differentiation of neuroblasts of the autonomic nervous system and of neuroectodermal mesen-chymal derivatives, using a biological cell marking technique. *Developmental Biology, 41*, 162–184.

Le Douarin, N. M., Teiller, M. A. M., & Le Douarin, G. H. (1975). Cholinergic differentiation of presumptive adrenergic neuroblats in interspecific chimeras after heterotopic transplantations. *Proceedings of the National Academy of Sciences U S A, 72*, 728–732.

McClune, B., & Jarman, R. (2012). Encouraging and equipping students to engage critically with science in the news: What can we learn from the literature? *Studies in Science Education, 48*(1), 1–49.

Mourez, M., Kane, R., Mogridge, J., Metallo, S., Deschatelets, P., Sellman, B., Whitesides, G., & Collier, R. (2001). Designing a polyvalent inhibitor of anthrax toxin. *Nature Biotechnology, 19*, 958–961.

Musheno, B. V., & Lawson, A. E. (1999). Effects of learning cycle and traditional text on comprehension of science concepts by students at differing reasoning levels. *Journal of Research in Science Teaching, 36*, 23–37.

Norris, S. P., & Phillips, L. M. (2008). Reading as inquiry. In R. A. Duschl & R. E. Grandy (Eds.), *Teaching scientific inquiry: Recommendations for research and implementation* (pp. 233–262). Rotterdam: Sense Publishers.

Norris, S. P., Stelnicki, N., & de Vries, G. (2012). Teaching mathematical biology in high school using adapted primary literature. *Research in Science Education, 42*(4), 633–649.

Nusslein-Volhard, C., Frohnhofer, H. G., & Lehmann, R. (1987). Determination of anteroposterior polarity in Drosophila. *Science, 238*, 1675–1681.

Phillips, L. M., Norris, S. P., & Macnab, J. S. (2010). *Visualization in mathematics, reading and science education*. Dordrecht: Springer.

Riddle, R. D., Johnson, R. L., Laufer, E., & Tabin, C. (1993). Sonic hedgehog mediates the polarizing activity of the ZPA. *Cell, 75*, 1401–1416.

Schwab, J. J. (1962). The teaching of science as enquiry. In J. J. Schwab & P. F. Brandwein (Eds.), *The teaching of science*. Cambridge: Harvard University Press.

Shepardson, D. P., & Pizzini, E. L. (1991). Questioning levels of junior high school science textbooks and their implications for learning textual information. *Science Education, 75*(6), 673–682.

Shkedi, A. (1998). Can the curriculum guide both emancipate and educate teachers? *Curriculum Inquiry, 28*, 209–229.

Shulman, L. S. (1986). Those who understand: Knowledge growth in teaching. *Educational Researcher, 15*, 4–14.

Swales, J. M. (2001). *Genre analysis: English in academic and research settings* (1990, 1st ed.). Cambridge, MA: Cambridge University Press.

Swales, J. M., & Feak, C. B. (2012). *Academic writing for graduate students: Essential tasks and skills* (3rd ed.). Ann Arbor: The University of Michigan Press.

Yarden, A., & Brill, G. (1999). The secrets of embryonic development: Study through research (In Hebrew and Arabic, 2004, 4th ed.). Rehovot: The Amos de-Shalit Center for Science Teaching.

Yarden, A., Brill, G., & Falk, H. (2001). Primary literature as a basis for a high-school biology curriculum. *Journal of Biological Education, 35*(4), 190–195.

Zer-Kavod, G., & Yarden, A. (2013). *Engineered bacteria produce biofuel from switchgrass* (an adapted primary literature article), Gene tamers – Studying biotechnology through research (In Hebrew, 2nd ed.). Rebovot: Department of Science Teaching, Weizmann Institute of Science.

Chapter 6
Teaching Scientific Reading

If adapted primary literature is to be a useful and important component of science education, then students must be capable of reading it. Such an ability is unlikely to be acquired automatically. Most students will require specific instruction designed to teach reading in a scientific context. Therefore, we need to know whether the instruction students receive currently in their schooling teaches them to read scientific text like APL, and, if not, what approaches might be taken to rectify the matter.

In this chapter, we will deal with three basic questions, the first two of which are: What do we know about the critical reading ability of young adults in the area of science? and, Does reading instruction in schools support the development of scientific literacy, characterized in this volume as being tightly linked to scientific inquiry? We document some evidence on the impoverished ability of young adults to read science. We then take another look at the idea of the simple view of reading introduced in Chap. 4 and show how it can help account for the state of affairs in science reading. We examine materials used for teaching reading in the early grades and ask about the extent to which such materials are designed to support the teaching of critical scientific literacy. Having reported that the guidance as it exists is poor, we turn to our third question: How might the teaching of reading in science be improved? We answer this third question by appealing to the insights suggested by an exemplary piece of science writing directed towards young children and by imagining how these insights might be imitated.

Young Adults' Critical Reading in Science

Although there are some very recent shocking statistics to the contrary (e.g., Shapiro 2013, March 13), a basic assumption of the education system is that high school graduates can read. Even if this assumption is false for too many students, it is certainly an aspiration of educational systems everywhere that high school

© Springer Science+Business Media Dordrecht 2015
A. Yarden et al., *Adapted Primary Literature*, Innovations in Science
Education and Technology 22, DOI 10.1007/978-94-017-9759-7_6

graduates should be able to read. In particular, from the perspective of many Western democracies, they should be able to read science for lifelong learning and participation in a democratic way of life. Even more particularly, high school graduates ought to be able to read media reports of science that appear in newspapers either of the traditional or electronic sort, in news magazines, or other magazines that carry science stories, and in internet-based sources generally. For example, high school graduates might be expected to read reports derivative from primary scientific literature that are relevant to making judgements concerning matters such as the relative strengths and weaknesses of medical treatments, the comparative costs and benefits of various means of home heating, the evidence on the use of garden pesticides, and the health benefits of choosing organic foods.

Another educational assumption is that learning to read takes place in primary and elementary school. However, we believe that school reading instruction needs to be examined critically because of how the idea of learning to read is constructed within schools. Typically, reading instruction takes place in the early years and normally developing children are assumed to know how to read by the time they are 10 or 11 years old. By that age, the thinking goes, they no longer need to learn to read and are ready to read to learn (Chall et al. 1990; Houck and Ross 2012). Coincidently, it is at this age that content-area instruction, such as science instruction, starts to predominate schooling. It often has been reported (e.g., Wellington and Osborne 2001) that content-area teachers, including science teachers, do not consider themselves reading teachers and take it for granted that the students they are teaching already know how to read (Wellington and Osborne 2001). A high school science teacher once asserted without defence to two of us, "I assign reading. I don't teach it."

In our judgement, young adults constitute a significant test group for the effectiveness of science reading instruction in schools. Young adults recently have completed or are nearing completion of compulsory schooling, and, as such, their reading competence provides a reasonable indicator of the success of schooling in teaching critical reading in science and of the levels of critical scientific reading to expect of non-scientist citizens in the future. Thus, if a person is nearing high school completion, or has a high school education, then the person should be able to read a science report found in a typical newspaper and prepared by Associated Press, say. Such reports usually are written at about a tenth grade reading level as assessed by such measures as the Flesch-Kincaid Grade Level score.

Following this line of thought, two of us asked senior high school students and university undergraduate students to read some media reports of science and to answer interpretive and metacognitive questions (Norris and Phillips 1994; Norris et al. 2003). We asked interpretive questions like these: For each of the following statements [taken from the reports and indicated to students], decide whether according to the report the statement is true, likely to be true, uncertain of truth status, likely to be false, false; and For each of the following statements [taken from the reports and indicated to students], decide whether the statement reports that one thing causes or influences another, that one thing is generally related to another, what was observed, what prompted the scientists to do the research, how the research was done.

The responses of the high school and university students were largely indistinguishable. The students: tended to demonstrate a certainty bias, rating statements more toward the "true" end of the scale than the statements were reported; had the most trouble interpreting the truth status of hedged statements; tended to interpret most accurately observation statements, statements of method, and predictions; tended to confuse causal and correlational statements; were liable to confuse descriptions of phenomena with explanations of them; and had difficulty distinguishing evidence from conclusions based on the evidence. In general, when the reading involved material that could be interpreted in isolation—facts about what was observed or done, statements about the future—the students performed not too badly. When the reading required integrating information from different parts of the texts and seeing connections between them, then they performed far less well.

For the university students we numbered the paragraphs in the reports and asked them to indicate in which paragraphs they found the information that helped them decide on their answers. The significance of this task is that completing it successfully relies on the location of information, which is at the nub of many school-based assessments. We found that about three-quarters of the students selected the paragraphs that were keyed correct. Coupled to this information-location task was a metacognitive question about how difficult they found the reports to read. At most, 5 % claimed that any report was difficult to read, and, in Goldilocks' fashion, more than 90 % judged the reading difficulty of all reports to be about right.

Let us put together these findings. Students knew the location of the information they needed to answer the questions. Students reported that the articles were not difficult to read. Yet, students did relatively poorly on the interpretive questions. We explain these results by appeal to the simple view of reading that has been shown to dominate instruction in reading (Collins Block and Pressley 2002; Pressley and Wharton-McDonald 1997). The simple view is that reading is being able to recognize the words correctly and to locate information in the text. If this is the view of reading you have, if the words in a text mostly are familiar to you, and if you believe you can locate the information needed to respond to questions you are asked, then you are prone to believe that the text is not difficult to read. This tendency will exist, even if you do not respond correctly to interpretive questions, because you will not have the means to make this negative assessment of yourself but only the means to make the positive assessment that the words are recognizable. The tendency also is exhibited by students who form contradictory beliefs after reading a text, by, for example, increasing their prior certainty in a point of view even though what they had read provided evidence against their view (Phillips and Norris 1999).

We are therefore forced to conclude that for a large swath of young adults their critical reading of science is deficient, and they are not being taught successfully to read in schools, if what is meant by reading is interpreting, analyzing, and critiquing texts. So what is to be done? We have a two-part answer. First, there needs to be a change in the conception of what it is to read, and, in particular, what it means to read science. Second, reading instruction in the early school years must have a component that addresses specifically scientific text, this instruction must not cease at the end of the early school years, and must be developed appreciably in the higher grades. One caveat is that by "scientific text" we do not have in mind typical science textbooks, but, rather, text that resembles APL in its similarity to primary scientific literature.

What Reading Is

The simple view of reading that pervades the educational establishment has had pernicious consequences. One of the foremost of these outcomes is giving rise to the assumption that ability to read is monolithic, as opposed to the view that reading ability not only exists on continua within different contexts but also can vary markedly between contexts. According to this alternative, multi-lithic, view of reading, competence in reading science can be expected to improve for many years even well into a professional career, and also to improve among non-scientists who choose to delve regularly into scientific materials. Also, according to the alternative view, the mere fact that a person can read well in one domain, say history, is no guarantee at all that the person can read well in science.

Given such considerations, it seems reasonable to question the makeup of a system that appears to assume that all anyone needs to know about reading can be learned in the first 10 years of life. The essence of the monolithic view of reading is that all reading is decoding, in the sense of figuring out what the words are. The view can make intelligible the difficulty faced by some children when they attempt to read the science materials typically found in school. Those materials are characterized by a dominant feature, namely, the presence of scientific vocabulary—terms that refer specifically to scientific concepts, processes, or methods (Norris et al. 2008). We acknowledge that it is to be expected that scientific text contain scientific vocabulary. However, that vocabulary does not have to be the dominant feature of those texts. As Chaps. 3 and 4 have shown, there are many other features of science upon which to focus attention. However, it is the combination of the simple view of reading and the dominance of scientific vocabulary in the sorts of scientific texts found in schools that leads to the view that science is difficult to read. This consequence arises because the simple view of reading directs students to focus their attention on decoding the words, and the text is brimming with unfamiliar words many of which pose exceptions to the rules for decoding more common words found in their stories.

In order to demonstrate concretely how the simple view of reading works, we often have turned to two contrasting examples. The first example mimics a typical school task designed to assess students' reading comprehension. There is a passage followed by a series of multiple-choice questions. Here is the task:

Seaview

During last night's severe winter storm the ten-story oil rig Seaview, operating on the Banks, capsized and sank. Rescue ships and aircraft are still in the area but no <u>survivors</u> have been found. The oil rig, with its 95 crew members, was operating for Petro Company in one of the richest fields in the world. It has been drilling since September 20. The pilot of one of the search and rescue aircraft reported that only debris could be seen on the surface. None of the rig's lifeboats were found. It is not known at this time what effect this disaster will have on Petro's future drilling program. There is as yet no official statement from the company.

Now, answer the five following questions:

1. The underlined word <u>survivors</u> means: A. people alive, B. people dead, C. lifeboats, D. life vests.
2. According to the article, the Seaview: A. turned over only, B. turned over and sank, C. was sinking, D. was nearly capsized.
3. According to the article, the future of Petro Company's drilling program: A. is bright and promising, B. is in serious difficulty, C. is unknown at this time, D. is certain at this time.
4. It can be inferred that the Seaview was drilling: A. on land, B. in the deepest part of the ocean, C. in a sheltered bay, D. in the ocean.
5. It can be concluded from the information in this passage that a probable cause of the sinking was: A. human error on the rig, B. the severe winter storm, C. poor rig construction, D. the huge size of the rig. (Norris and Phillips 2008, p. 244–245)

Check your answers against ours, which are: ABCDB. This type of task is a familiar sort found in school. If students score well on tasks of this sort, the conclusion that is often inferred is that the students understood the passage. It is our contention that this inference is not justified and therefore that the conclusion cannot be trusted. A corollary of our contention is that students can do well on a test of this sort without understanding at all beyond word identification and grasping the syntax. We sometimes have found it difficult to convince teachers of our point, because they understand the passage and they get the items correct. They thus see as evidence against our view the correlation between their performance and their grasp of the passage. How then can we support the claim that getting the items correct can actually be due to processes that do not go beyond surface understanding and provides little or no evidence of a level of understanding that is educationally significant? Our usual approach has been to ask the teachers to read another passage and to answer a set of multiple-choice questions based upon it.

Quantum Damping

We assumed that the atomic energy levels were infinitely sharp whereas we know from experiment that the observed emission and absorption lines have a finite width. There are many interactions which may broaden an atomic line, but the most fundamental one is the reaction of the radiation field on the atom. That is, when an atom <u>decays</u> spontaneously from an excited state radiatively, it emits a quantum of energy into the radiation field. This radiation may be reabsorbed by the atom. The reaction of the field on the atom gives the atom a linewidth and causes the original level to be shifted. This is the source of the natural linewidth and the Lamb shift. (Louisell 1973, p. 285)

The questions are as follows:

1. The underlined word <u>decays</u> means: A. splits apart, B. grows smaller, C. gives off energy, D. disappears.
2. According to the article, observed emission lines are: A. infinitely sharp, B. of different widths, C. of finite width, D. the same width as absorption lines.
3. According to the article, the most fundamental interaction that may broaden an atomic line is: A. the Lamb shift, B. the action of the atom on the radiation field, C. the emission of a quantum of energy, D. the reaction of the radiation field on the atom.
4. It can be inferred that when an atom decays it may: A. return only to a state more excited than the original one, B. not return to its original excited state, C. return to its original excited state, D. return to a state less excited than the original one.
5. It can be concluded from the information in this article that the assumption that atomic energy levels are infinitely sharp is: A. probably false, B. false, C. true, D. still under question. (Norris and Phillips 2008, p. 245–246)

Check your answers against ours: CCDCB. You will have noticed that the format of the second passage and accompanying questions is precisely the same as that of the first. There is a word meaning question, literal interpretation questions (items 2 and 3), and inferential interpretation questions (items 4 and 5). We have given these passages to hundreds of students and teachers and we have detected no discernible difference between the level of performance on the two tasks. Yet, with the very few exceptions of individuals who have studied quantum physics, they have readily admitted that they do not understand the Quantum Damping passage at all beyond being able to identify most of the words and decipher the syntax.

So why are people able to perform so well on items about a passage that is totally beyond their comprehension? Our claim is that for items such as those used, word and syntax identification are sufficient for good performance. If this is true for the Quantum Damping passage, then it also must be true for the Seaview passage. The logical consequence is that tests such as these do not provide evidence of the level of understanding that science educators try to promote in their students. Unfortunately, tests such as these find widespread use in schools and provide for teachers and policy makers the dangerous illusion that students are doing better than they are.

If the critical reading of science is to replace the simple view of reading as word recognition, then science education will need to face two discomfiting truths. The first truth is that on average high school graduates cannot read science, or cannot read science very well, if reading is taken to mean critical reading. The second is that students are not taught to read science in school, and indeed little instruction in reading scientific texts occurs either in science or reading lessons (Heselden and Staples 2002; Palincsar and Magnusson 2001; Shymansky et al. 1991). In the following section, we take up the issue of how these disagreeable truths might be confronted.

Teaching Science Reading in School

We begin by describing science reading instruction in elementary school, specifically in grades one to six, which include children from 6 years old to about 12 years old. Most reading instruction in school takes place at this level. Our focus will be on reading instruction that is guided by the use of commercial reading programs, which are used commonly in North America and elsewhere. We will point to various shortcomings found in those programs. Then we will offer some suggestions about how science reading could be taught differently, both in elementary school and in high school. There remains room for reading instruction in science beyond even high school, but we will not address that age range here.

Science Reading Instruction in Elementary School

We know that many elementary teachers rely on commercial reading programs for reading instruction (Morrow and Gambrell 2000; Moss and Newton 2002; Smith et al. 2004). So it is reasonable to ask about the extent to which commercial reading programs are designed to support the teaching of scientific reading. In an extensive study of three such programs Norris et al. (2008) provided some answers to this question. We presume those answers provide insight into how reading in science is taught in the early grades even by teachers who do not use commercial reading programs, although we have no direct evidence on those other teachers.

Each reading program we examined is based upon a compilation of shorter works selected from many sources and meant to satisfy a diversity of interests and reading levels. About 22 % of the selections in the three reading programs contained science and technology content. The programs range from grade 1 to grade 6, and, although there was some variation, science selections were included for each grade level. So, there is a significant amount of science content in these programs, which on the face of it is good news for science education. Also significant, about two-thirds of the selections that contained science content were almost entirely science. There was a marked difference in emphasis of science areas, with about 40 % of the selections falling into the life sciences and fewer than 5 % into the physical sciences.

What turns out to be a very significant finding for the teaching of reading in science is the way in which the genre of the selections was distributed between science and non-science selections. Whereas, about two-thirds of the non-science selections were literary (either narrative or poetry) in genre and fewer than 10 % expository, more than one-half of the science selections were expository and only 16 % of them literary. We will come back to the significance of this finding very shortly after we have reported on the text features and the instructional guidance recommended to teachers.

The frequency of types of text features in the science selections was also a distinctive characteristic. By far the most dominant features were photos and scientific vocabulary, which were represented about equally in numbers and which counted for nearly one-half of all the text features documented. In contrast, charts, graphs, tables, and scientific meta-language, which are integral pillars of primary scientific literature, were present hardly at all.

The key to understanding the importance of these data lies in the nature of the instructional guidance offered to teachers. Notably, traditional "skill and drill" and information location and word identification were not among the primary guidance offered. Given the prominence of the simple view of reading, this finding is important because reading programs have been criticized historically for promoting skill-based instruction. The key question for us is: What kinds of literacy instructional guidance was associated with the selections that contained science content? Although, a multitude of types of instructional guidance appear, the directives and questions that were suggested most frequently were those that involve personal reflection and response. Here are some examples: Will this report change the way you use water? Why do you think astronomers study the stars? What did you learn about water that surprised you? Did you like the story? Have you ever used a compass to find your direction in an unfamiliar place?

Our complaint is not that such questions are invalid, unimportant, or mis-educative. Rather, it is that these sorts of questions represent almost all of the questions that were provided as guidance to teachers. None of these questions is what we would call a scientific question. All of these questions share an association with the personal—involving personal experiences, preferences, learnings, or ideas. Again, we recognize that such questions can lead to scientific questions, but in the instructional guidance offered to teachers they did not.

How Science Reading Instruction Might Be Different in Elementary School

Perhaps the most scientific of all the selections we identified in the commercial reading programs is one suitable for fourth or fifth graders. Entitled *Dancing Bees,* written by Margery Facklam and illustrated by Pat Stephens (Facklam 1994), the book is based upon the idea that bees perform a sort of dance to communicate about food sources and describes an ingenious experiment used to find this out. The potential of this selection for teaching reading in science is not realized because the instructional guidance focuses on students' personal responses, such as sharing personal interpretations of dance, developing their own dances, and imitating the bees' waggle dances. Seeking such responses is fine, as we have said already, but it does not teach the students about reading science.

The selection opens with the description of a phenomenon: namely, that honey-bees who find a supply of nectar seem to have a way of communicating its location to the other bees in the hive. The selection then raises the scientific question: "How can an insect with a brain no bigger than a grass seed pass on all this information?" That question can serve as the basis for a lesson on science reading. First, students need to be taught to recognize the question as scientific. That is, the question is neither rhetorical, nor a request for an opinion nor does it seek information that is already known. The question might be an invitation for a conjectured answer, but the key to the lesson for the students is that this is a question that must be answered scientifically, that is, through the results of a scientific investigation. Seeing the question in this light is quite a contrast to how questions in literary texts are seen.

The selection continues to describe the conclusions made by Dr. Karl von Frish based on observations of thousands of bees. Bees who have located a source of nectar return to the hive, whir their wings, dance in a figure eight, and provide samples of the newly found food. The bees traverse the line between the top and bottom half of the eight while travelling in the same direction whether completing the top or bottom half. That direction points to the nectar. The distance to the nectar is indicated by dancing speed: the farther away the faster the dance.

The book continues by describing an experiment conducted by a team of scientists from Denmark and Germany to test the accuracy of the above conclusions. Peppermint-scented sugar water was placed as a food source about 1.5 km from a bee hive. When the food was found by a worker bee, the bee returned to the hive, provided instructions in the usual way, which almost 300 bees interpreted correctly and located the food source themselves. Next, the scientists moved the food source and introduced a computerized robot-bee into the hive. The robot had been programmed to perform a dance in a manner that the scientists believed would communicate correctly the new location of the food and to provide a sample of the food itself. The book reports that 100 bees found the sugar water compared to about 10 bees when any part of the performance was omitted—the dance, the sample of food, or the whirring wings.

Dancing Bees thus contains the central ingredients of primary scientific literature: a research question, a hypothesized explanation of a phenomenon, a description of methodology, reported results, and a conclusion. Although instruction in reading would need to attend to the individual words that might be unfamiliar to students, to teach reading from a scientific point of view there needs to be a focus on these central ingredients. In the first place, students need to be taught to differentiate between the various ingredients: Here is a research question; this section describes the research methodology; the results of the experiment are being reported now; and so on. Second, the students require instruction on the critical stance that it is appropriate to take toward each of these ingredients. For example, the research question might be analyzed for its clarity and its susceptibility to investigation. The methodology can be examined from the point of view of its appropriateness, practicability, and likely effectiveness in answering the research question. The conclusions can be queried for their justification in light of the results; whether they have overreached the evidence, or not reached as far as they might.

Obviously, beginning students cannot be expected to cope easily or sophisticatedly with matters such as these. They require practice based upon many opportunities with a variety of texts in order to develop the critical reading ability envisaged. They need to hear the questions being raised, to see answers to them modelled by the teacher, and to hear reflection on the answers. This instructional process must extend over all of the elementary years, beginning in the earliest years with very simple texts and with verbal discussions about scientific activities that raise these issues in age-appropriate terms.

Reading taught in the manner just described calls for teachers who are both comfortable with science and are aware of its basic epistemology. Comfort with science is required because the issues raised will not always be straightforward. For example, it will not be clear in all cases that the research methodology is suitable for answering the questions asked and that the conclusions follow from the evidence. Yet, it is just these sorts of ambiguities that point to the epistemology of science. As we have described it in Chap. 3, that epistemology is both rational and fallible. The fallibility points to the liability for error and the attendant possibility of needed change. The rationality points to the dependence upon reasons for all decisions that are made. It takes a confident and astute teacher to deal with situations marked by such a possibility of fluidity. The traditional view that science is a body of determined and stable knowledge is soon replaced by one that is dynamic and forever in flux.

How Science Reading Instruction Might Be Different in High School

We should say from the outset that any science reading instruction in high school would be different than what occurs in most high school science classes at present, because little such instruction exists in them. Instruction that does take place typically is centred on the learning of new vocabulary, of which there can be an extensive amount in some courses. Scientific vocabulary is important because without it there can be no deep understanding of the substantive content of science. On the other hand, vocabulary that refers to the substantive content of science is quickly forgotten by those who do not use it frequently, that is, those who form the great majority of high school graduates.

The alternative to focussing excessively on scientific vocabulary is to attend to those aspects of scientific language that might be easier to recall, and also more useful to the experience with science had by those who do not make science a career, but who still aspire to be scientifically literate citizens. We established in Chap. 4 that reading science is a proper goal of science education. In making a case for that conclusion, very little reliance was placed upon learning scientific vocabulary. Rather, the reasons why reading science is a proper goal of science education have to do with the access that reading provides to the nature of scientific inquiry and thence to critical scientific literacy, and with the fact that reading science is actually integral to scientific inquiry itself making the teaching of reading part and parcel of teaching scientific inquiry.

One productive way to proceed with science reading instruction at the high school level would be to focus on the features of scientific text identified in Chap. 3: text structure, text epistemology, and meta-scientific language. Attention to each of these features can guide a reader to sounder interpretations of scientific text than otherwise would be possible.

Text Structure

Recall that three types of text structure were identified and illustrated: organizational, goal-directed, and argumentative. Organizational structure provides the reader first with a general sense of location in the text (e.g., objectives section, methods, results, discussion, etc.) and of the speech acts to be expecting (e.g., reporting if in the methods and results sections; inferring if in the conclusions section; possibly speculating if in the discussion section). Teaching students how to identify these structural signals in text and how to adapt their stances accordingly is fundamental to learning to read scientific text.

Goal-directed structure is more fine-grained than organizational structure. Students must learn to recognize both overall goals, those proposed for the entire study being reported, and more local goals, such as the purpose for taking the measurement of a particular parameter. To miss the goal-directed structure of the text is to miss the purpose for the writing and hence to lose the point of the text.

Argumentative structure also can be an organizational pattern for the text as a whole or be revealed at the section or paragraph levels. Considering the text as a whole, the argumentative structure often links the main conclusion to the primary evidence reported (if the text is reporting a data-based study). At finer detail, the methods section often presents an argument that defends the adoption of the methods followed; the introductory section might provide reasons for conducting the study; the discussion might attempt to discredit alternative interpretations of the data than the one favoured by the authors. Each of these interpretive challenges is a potential focus for reading instruction in the high school science curriculum.

Text Epistemology

Interpreting the epistemology contained in a scientific text comes down to identifying expressions of fallible rationality. Rationality, as we discussed in Chap. 3, is made concrete and perceptible through reasons and evidence offered in support of conclusions. Outward evidence of fallibility is constituted through the use of modal auxiliaries such as "may" and "might" and adverbs such as "perhaps", "possibly", and "conceivably". It is impossible to understand a scientific text without understanding the epistemology expressed or implicit in it. Thus, students must be taught the multiple ways that text can be used to express both rationality and fallibility.

Achieving an ease of interpretation with text epistemology is not likely to come easily or quickly, and likely will continue to develop after high school for those who pursue college and university educations. Nevertheless, learning to interpret text epistemology is of sufficiently significant consequence for scientific literacy that considerable effort should be expended on it in the high school science curriculum.

Meta-scientific Language

Interpreting meta-scientific language takes the reader once more to the granular level of the text. Although we have played down the importance of scientific vocabulary for scientific literacy, we believe that understanding the functions of scientific meta-language is very important. Much of the importance resides in the fact that meta-language use is related closely to expressions of epistemology. For example, use of the word "observation" indicates that the scientist is reporting something that for the nonce will be taken as given. Use of the word "hypothesis", on the other hand, signals quite differently that the scientist is announcing something that is conjectural, open to test, and possibly will be tested in the study being described. If students do not learn, first, to make distinctions such as these, and, second, to notice the use of meta-scientific language when they are reading scientific text, they will not be able to grasp the meanings of those texts. Once more, direct instruction, modeling by the science teacher, and plenty of practice are the ingredients for success by teaching this aspect of reading in science.

Conclusion

We conclude this chapter with a set of take-home messages (T-hM). These are ideas for teachers and about teachers that concern the teaching of reading in science. Some of the messages are cautionary; some are exhortations. Taken together they can form the basis for a renewed pre-service and in-service teacher education focused on reading in science.

T-hM-1: Reading Programs Provide No Reliable Direction to Teachers on which Selections Are Science

It is important for teachers to know and to keep in mind this fact about reading programs. It is a cautionary message that alerts teachers to the fact that the task of identifying scientific selections falls to them. The message is equally important for elementary school teachers, who use reading programs to teach, and secondary school teachers, who tend to believe that the teaching of reading falls to those who had the students before them. The message tells those secondary teachers that elementary teachers operate under a serious handicap that can affect negatively the background in reading science that their students have had.

You might ask what difference a categorization into science and non-science selections might make to the teaching of reading. Imagine a poem about the variety of animal tails (Cornerstones 2a, 2000). The poem has five stanzas of four lines each with a rhyming pattern a-b-a-b. The opening stanza describes some variation in tail appearances—long, short, bushy, curly. The following three stanzas speak of uses of tails—for swinging, wagging, warning, balancing, swatting, showing off, and steering. It is a fun poem and easily could be treated solely as a literary genre and analyzed from that perspective. Yet, the poem refers to several what in science are called "functional explanations". So, there is teachable science also evoked by this poem. What tail functions are mentioned? What other functions might tails have? What are the functions of other body parts? What is the nature of functional explanations? Do functional explanations really explain? Do functional explanations presuppose a designer? If the functions of body parts evolve, does evolution also presuppose a purpose or a direction? Clearly, it makes a difference to the lesson whether the teacher chooses to categorize the selection as literary or scientific, or both.

T-hM-2: How Teachers Perceive and Understand the Text Affects How They Teach Reading in Science

The message here is over and above the previous message about identifying selections as having scientific content. In elementary school reading programs it is possible to have several selections, all with scientific content, but having different genres. One might be an expository text, another a poem such as the one about tails, and the other a narrative, for example. In the case of the expository genre, there is a temptation to focus upon the factual content. In so doing, two risks emerge. The first is the promotion of the simple view of reading as information location. The second is the further inculcation of the widespread view of science as a collection of disparate facts. Deep and sophisticated understandings of both reading and science are needed for a teacher to navigate successfully around both of these risks.

In the cases of the narrative and the poetry the temptation is to focus on the literary qualities and to overlook the science. This temptation might be particularly enticing for teachers whose science background is not strong, as is the case with most graduates of elementary school teacher education programs. Instructional guidance that encourages teachers to elicit personal experiences and reflections from students and that does not encourage attention to the scientific writing only exacerbates this problem.

Special background is needed by teachers to achieve an effective scientific focus when a text like *Dancing Bees* is included among reading selections. As we have said, this text contains truly scientific writing. Unless these qualities are recognized by teachers, then the potential power of a text of this nature will be lost. Furthermore, given that *Dancing Bees* is exceptional, and that the majority of selections with science content are written as expositions, there is a further tendency to associate science with fact telling, a connection that perhaps only teachers with astute science sensibilities would notice and take measures to counteract.

T-hM-3: Children's Perceptions of Science Are Affected by the Content Focus Chosen by the Teacher of Science Reading

Even within science selections there are risks of misrepresentation. We have reported that among all of the science selections in the three reading programs we studied, only a very small minority fell into the physical sciences and most fell into the life sciences. Understand that our preference always will be to opt for a science selection, whatever the content area, so long as few science selections exist overall. However, in the commercial reading programs we surveyed about one-quarter of the selections contained scientific content, which in our judgment is a substantial and satisfactory proportion. Given that the overall proportion of science selections is adequate, we can turn attention to the representation within the sciences and we find that the proportion of physical science, at less than 5 %, is not acceptable. Not only is knowledge from the physical sciences important and often needed to grasp scientific ideas in the life sciences, the neglect of the physical sciences risks calibrating young students' interests so that by default they favor life sciences over physical sciences and playing into a pre-existing partiality of elementary school teachers for the life sciences over physical sciences (Appleton 2003).

T-hM-4: The Meta-Language that Structures Scientific Writing and that Reveals Scientific Reasoning Should Be an Object of Instruction in Reading Science

The particular uses of meta-language help set scientific writing apart from other writing. Scientific meta-language provides the concepts for talking about science, and it is an ability to talk intelligently about science that is the mark of scientific literacy. Paying attention to even a very small subset of this meta-language can have an enormous difference. Hand and his colleagues (e.g., Nam et al. 2011) have shown, for example, that teaching students to use only the three concepts—"scientific question", "claim", and "reason"—can lead to considerable improvement in young students' knowledge of and facility with scientific reasoning, reading, and writing. With a larger repertoire of language, individuals have the facility to talk about the causes and effects explored in a study; to notice what was hypothesized, what was observed, and what were the results; and to distinguish between data and explanations. With a supply of concepts such as these, individuals are able both to detect the reasoning in scientific writings and to reason about the writings themselves.

T-hM-5: Every K-8 Teacher and Every Science Teacher Should Be a Teacher of Science Reading

We showed in Chap. 3 how very important reading is to scientists. In Chap. 4 we argued that reading must be important to science education as well. We provided several grounds for this conclusion. First, reading scientific text is part of scientific inquiry and thus has a legitimate call for being included in any science curriculum that aims to teach scientific inquiry. Second, reading science is the fundamental sense of scientific literacy, and scientific literacy is widely understood as the most important goal of science education. Third, any robust sense of reading must include critical reading and critical reading bolsters the type of critical scientific literacy that is most desirable. Finally, because reading is a genuine form of inquiry, reading science provides access to specimens of scientific inquiry recorded in scientific writings.

Given its importance to science education, no additional argument is needed to conclude that science teachers should teach reading in science. The case for elementary school teachers rests on the fact that the foundation for good reading is laid in the early years of schooling, particularly the first 5 years. Much of past practice in teaching reading seems to presuppose that reading is monolithic and that competence in reading stories, for example, is sufficient to read other genres, including science. We have shown that this presupposition is no longer tenable and that specific instruction in teaching reading in science is necessary. Given that in elementary schooling subject-matter specialist teachers are rare, then the task of teaching science reading falls to the classroom teachers.

References

Appleton, K. (2003). How do beginning primary school teachers cope with science? Toward an understanding of science teaching practice. *Research in Science Education, 33*, 1–25.

Chall, J., Jacobs, V., & Baldwin, L. (1990). *The reading crisis: Why poor children fall behind*. Cambridge, MA: Harvard University Press.

Collins Block, C., & Pressley, M. (2002). *Comprehension instruction: Research-based best practices*. New York: Guilford.

Facklam, M. (1994). *Big bug book*. Boston: Little Brown & Co.

Heselden, R., & Staples, R. (2002). Science teaching and literacy, part 2: Reading. *School Science Review, 83*, 51–62.

Houck, B. D. & Ross, K. (2012). Dismantling the myth of learning to read and reading to learn. *ASCD Express, 1*(11). Retrieved October 4, 2013, from http://www.ascd.org/ascd-express/vol7/711-houck.aspx

Louisell, W. H. (1973). *Quantum statistical properties of radiation*. New York: Wiley.

Morrow, L. M., & Gambrell, L. B. (2000). Literature-based reading instruction. In M. L. Kamil, P. B. Mosenthal, P. D. Pearson, & R. Barr (Eds.), *Handbook of reading research* (Vol. 3, pp. 563–586). Mahwah: Lawrence Erlbaum.

Moss, B., & Newton, E. (2002). An examination of the informational text genre in basal readers. *Reading Psychology, 23*, 1–13.

Nam, J., Choi, A., & Hand, B. (2011). Implementation of the science writing heuristic (SWH) approach in 8th grade science classrooms. *International Journal of Science and Mathematics Education, 9*, 1111–1133.

Norris, S. P., & Phillips, L. M. (1994). Interpreting pragmatic meaning when reading popular reports of science. *Journal of Research in Science Teaching, 31*(9), 947–967.

Norris, S. P., & Phillips, L. M. (2008). Reading as inquiry. In R. A. Duschl & R. E. Grandy (Eds.), *Teaching scientific inquiry: Recommendations for research and implementation* (pp. 233–262). Rotterdam: Sense Publishers.

Norris, S. P., Phillips, L. M., & Korpan, C. A. (2003). University students' interpretation of media reports of science and its relationship to background knowledge, interest, and reading difficulty. *Public Understanding of Science, 12*, 123–145.

Norris, S. P., Phillips, L. M., Smith, M. L., Guilbert, S. M., Stange, D. M., Baker, J. J., & Weber, A. C. (2008). Learning to read scientific text: Do elementary school commercial reading programs help? *Science Education, 92*, 765–798.

Palincsar, A. S., & Magnusson, S. J. (2001). The interplay of first-hand and second-hand investigations to model and support the development of scientific knowledge and reasoning. In S. Carver & D. Klahr (Eds.), *Cognition and instruction: Twenty-five years of progress* (pp. 151–194). Mahwah: Lawrence Erlbaum Associates Inc.

Phillips, L. M., & Norris, S. P. (1999). Interpreting popular reports of science: What happens when the reader's world meets the world on paper? *International Journal of Science Education, 21*(3), 317–327.

Pressley, M., & Wharton-McDonald, R. (1997). Skilled comprehension and its development through instruction. *School Psychology Review, 26*, 448–467.

Shapiro, B. (2013, March 13). 80 % of NYC HS grads entering City College don't have basic skills. *Breitbart*. Retrieved October 15, 2013, from http://www.breitbart.com/Big-Govern ment/2013/03/07/NYC-80-percent-read

Shymansky, J. A., Yore, L. D., & Good, R. (1991). Elementary school teachers' beliefs about and perceptions of elementary school science, science reading, science textbooks, and supportive instructional factors. *Journal of Research in Science Teaching, 28*(5), 437–454.

Smith, M.L., Phillips, L.M., Leithead, M.R., & Norris, S.P. (2004). Story and illustration reconstituted: Children's literature in Canadian reading programs. *Alberta Journal of Educational Research, 50*(4), 391–410.

Wellington, J., & Osborne, J. (2001). *Language and literacy in science education*. Buckingham: Open University Press.

Chapter 7
Applications of Adapted Primary Literature

In this chapter we discuss the outcomes of studies that examined the applications of Adapted Primary Literature (APL) for reading and learning science presented in Chaps. 5 and 6. In Chap. 5 we discussed the procedures to create Adapted Primary Literature (APL) and how to use it for teaching science in schools, and in Chap. 6 we discussed the approaches to teach scientific reading in schools.

According to the science framework of the Program for International Students Assessment (PISA, OECD 2012), scientific literacy requires not only knowledge of the concepts and theories of science but also a knowledge of the common procedures and practices associated with scientific inquiry and how these enable science to advance. It is claimed in the framework that individuals who are scientifically literate should have knowledge of the major conceptions and ideas that form the foundation of scientific and technological thought; how such knowledge has been derived; and the degree to which such knowledge is justified by evidence or theoretical explanations.

We hypothesized that if high-school students were able to read APL articles, they could probably develop the following components of scientific literacy: (i) acquaintance with the rationale of the research plan and procedures; (ii) exposure to research methods and their suitability to the research question; (iii) acquaintance with the language and structure of scientific communication; (iv) development of the ability to critically assert the goals and conclusions of the scientific research; (v) exposure to problems in a certain discipline, and the ways scientists try to solve them; and (vi) acquaintance with the continuation of the scientific research process: the results of a certain research report may answer the initial research question, but may also raise a new research question, to be answered by additional research. In addition, we contemplated that students may find reading research articles a novelty and a challenge, and may also identify with the quest of the researchers reporting their work (Yarden et al. 2001). Since then several seminal research studies have attempted to examine these hypotheses. We outline next the outcomes of those studies, with an attempt to demonstrate the benefits as well as the challenges involved in the use of APL for science teaching and learning.

© Springer Science+Business Media Dordrecht 2015
A. Yarden et al., *Adapted Primary Literature*, Innovations in Science
Education and Technology 22, DOI 10.1007/978-94-017-9759-7_7

We describe the outcomes of using APL alone, and the outcomes of using APL in conjunction with or in comparison to other scientific text genres, namely Primary Scientific Literature (PSL), Secondary Literature (or Journalistic Reported Version, JRV), or textbook. Each section will start with a description of the use of these scientific text genres in laboratory settings and then continue with its use in science classrooms.

Outcomes of Learning Science Using APL

Previous research on reading has revealed that learning from texts is a complex skill, which involves complex interactions between a reader's relevant prior knowledge and the relevant information in a text, rather than a one-way flow of information from the writer (text) to the reader and vice-versa (Holliday et al. 1994). Linking the content of the text with existing knowledge structures (Kintsch 1989), constructing meaning (Samuels 1983), and being able to represent knowledge in different ways, while maintaining motivation and interest (Asher 1980; Epstein 1970) are only a few examples of the complexity involved. Expert scientists perform these cognitive actions automatically when they read scientific texts because they have extensive specific experiences and knowledge, however novice readers may find it a difficult task (Alexander and Jetton 2000; Samuels 1983) because their experiences and knowledge are limited by comparison. Thus, beyond the complexity involved in reading per se, reading scientific texts adds an additional complexity for novices who are neither familiar with the genre nor the structure of such texts. In this section we describe a study that examined the outcomes of reading APL in a laboratory setting (without teacher intervention) followed by studies that examined the outcomes of the use of APL in science classrooms.

Learning an APL Article in a Laboratory Setting (No Teacher Intervention)

Using a qualitative approach, we analyzed the way two 12th grade students read an APL article from the curriculum in developmental biology in a laboratory setting without teacher intervention (Brill et al. 2004). In contrast to expert readers of scientific texts who usually glimpse through the text and attend only to parts of it, the two students read the entire article from beginning to end. Although questions for the APL article were handed to the students with the article, the students ignored the questions they were to answer and first read the article. We could therefore characterize two main stages in the reading process of the two students through the APL article: initial reading in which the readers reported they fully understand the article (because they could "read" the words), and a second stage in which the students realized their miscomprehensions through monitoring what they read and attempted to rectify the misconstrued details.

The two students had a feeling that they understood the article after reading it for the first time. In contrast, during the second stage of reading, the two students understood that there was a deeper level to the text and that their understanding had been superficial in their first reading of the article. The students first read of just the article without attention to the questions exposed their approach and enabled us to observe the naive way in which they constructed knowledge from the research article, without the aid of any directing questions. One important problem was that these students did not distinguish between multiple experiments presented in the article. Instead of a fragmented view of the article, which might be expected from novice readers, they constructed their own meaning from the text, connected partial information pieces in the article (an experiment described in the introduction section and an experiment from the results section) that were separate and distinct experiments. In addition, since they misunderstood the role of the experiments described in the previous sections (introduction, methods) they did not read the results section of the article thoroughly and thus missed the main point of the research.

An example of differences between expert and novice readers and their comprehension of the research article readers can be seen in the way the two students understood the abstract. One of them described difficulties in understanding the abstract, claiming that she could not put the details together and create a coherent understanding of the text. In contrast with this fragmented cognitive structure, expert readers construct a schema of the main outline of the research to be read based on the abstract, although they have never read the specific research before because they have knowledge of the often minimal details in an abstract. The abstract thus serves as a form of advance organizer for the expert readers because of their advanced knowledge structures and familiarity with the purpose of an abstract (Ausubel 1963). In contrast, these students who are novice readers, could not construct meaning from the abstract, and may have read other sections of the APL article (for example, the introduction or the methods) as separate and distinct even though there was similarity in form between the experiments described as well as the results. Indeed, the two students regarded the information in the introduction as information about the research itself, and could not distinguish between the experiments described there and in the results section.

It is reasonable to expect that dividing scientific articles into distinct parts—which summarize the article (abstract), give the relevant scientific background (introduction), report the scientific methods (methods) and the results of the research (results), and finally discuss their meaning and raise new research questions (discussion) would promote a consistent and complete comprehension of the information provided by the research. Without any doubt, the expert reader has a firm concept of what the different parts of the article should entail. These conceptual abilities, and the orderly way in which the scientific research is reported in PSL articles, enable expert readers to read, comprehend and even criticize the research (Alexander and Jetton 2000; Yarden et al. 2001). The different sections of PSL and APL articles can be considered informational elements that have no meaning when read separately and therefore may have contributed to cognitive overload for the two novice readers. Reading the methods section for the first time may have had limited meaning for the two students without an understanding of the rationale for and the consequent results of the research

The evidence of how these two novice readers dealt with the APL articles provides a window into how they attempted to comprehend the article. They read the APL article in a chronological manner and it seems reading the methods section before the results section had limited meaning to them. To overcome this problem—namely, in order to give meaning to the methods section—the two students started thinking in the methods section about the experiments and their predicted results. Subsequently, when they reached the results section, they were aware that they already knew the experiments and had even predicted their results. Therefore, they did not read the most important section of the article (the results section) thoroughly and consequently could neither distinguish between the experiments, nor relate the relevant method to each corresponding experiment. The same problem might have occurred when they read the results section. Without the interpretation, the results had no meaning for the two students. One of the students indicated the cognitive load she encountered when reading the article: 'the fact that the results and the explanation of the results are not together is quite confusing. Because, I mean . . . you read the results and then you say: OK, what does it mean? . . . and then when you read the discussion you have to go back to the results again because you don't remember what was there'. She then recommended that a student reading the article for the first time, read the introduction section and the discussion and only then the methods and results. Indeed, Sweller and Chandler (1994) suggested that one reason for cognitive overload may be the physical separation between the related pieces of information in the material to be learned.

The two students could not distinguish between multiple experiments that appeared in the article. A possible reason for this difficulty is the fact that the method used in the experiments may be identical, although the research question is different. The novice readers of APL articles that we followed (both in laboratory settings and in classrooms) needed guidance in distinguishing between what seemed to them to be identical experiments. Alexander and Jetton (2000) suggested a developmental model of learning from text, which involves three levels of academic development: acclimation, competence, and expertise. The way in which the two students read the APL article for the first time indicated, at times, that they were in the acclimation (or initial) stage, and at times that they were at the competence (or intermediate) stage. Their relevant prior knowledge was limited, but they seemed to be able to overcome some coherence gaps to reach a superficial understanding without teacher help. In order to learn from APL articles, readers must be flexible, construct tentative schemas, and monitor their fit with relevant information in order to construct more comprehensive and complete interpretations and it appears that the corresponding questions to the research article may prompt students towards this advanced stage of reading. In addition, to facilitate the transition from the initial phases of learning to the intermediate phases, our curriculum includes a well-structured and coherent introduction to the specific scientific domain (i.e., embryonic development, Yarden et al. 2001).

Learning an APL Article in a Classroom Setting

Our experience with learning through APL in classrooms is limited to the use of the APL-based curriculum for 12th grade biology majors class (Falk et al. 2008; Falk 2009; Falk and Yarden 2009), and to the use of a web-based APL prototype in mathematical biology (Norris et al. 2009). We outline our major findings from these three classroom settings next.

Using a constructivist qualitative research approach, we examined the benefits and challenges to students learning through an APL-based curriculum in biotechnology (Falk et al. 2003) reported by both the students and their teachers and reflected in video-recorded class episodes (Falk et al. 2008; Falk 2009). We found that the APL curriculum promotes *engagement, knowledge integration, inquiry thinking, discipline-specific epistemic beliefs*, and increased *comprehension of subject matter* among the high-school students. The main challenge we identified was comprehension of biotechnology processes and methods that require the use of prior knowledge in new contexts which often led to cognitive overload. We provide some details and supports for these findings below.

Students' Cognitive and Affective Engagement Students exhibited cognitive and affective engagement during the enactment of the APL-based curriculum in biotechnology in the four classes investigated while learning the curriculum. Students' engagement was strongly cherished by the enacting teachers, because students continuously provided positive feedback. During the students' group interviews, the most prevalent category of positive responses referred to their engagement during the study of the biotechnology curriculum. Several students reported that they were more interested in reading a research article than a textbook or the Introductory Unit of the curriculum and even suggested using articles rather than textbooks in class because they elicit more interest. Students attributed their engagement to the contemporary nature of the curriculum content, to the ability to solve real-world problems using school science knowledge, and to the authenticity of this curriculum genre in contrast to common biology curricula. Although the APL-based curriculum in biotechnology is more complex than the other adapted articles because of the many molecular processes it presented, the article that raised the highest level of interest was the gene-therapy article, which deals with human subjects.

Knowledge Integration In contrast to other curricula where the teachers perform the integration and present it to the students, teachers reported that the use of APL prompted them to direct their students to actively integrate their relevant prior knowledge with the relevant text knowledge by themselves. The teachers reported that by suitable reference to previously learned subjects and through the contents of the Introductory Unit to the curriculum, it was possible to facilitate the integration process.

Inquiry Thinking In the context of learning using the APL-based curriculum, one of the major points teachers focused upon in the interviews was students' acquisition of inquiry-thinking skills while learning APL. The fact that both students and teachers said that the curriculum provides opportunities for *learning by inquiry* is not a trivial one, since it might be expected that learning using APL will mainly enhance *learning on inquiry* (the terms learning by inquiry and learning on inquiry are used here following Tamir 1985). Teachers and students explicitly attributed any gain to the APL genre and to the problem-solving epistemology of biotechnology reflected in the curriculum. It seems that APL conveys a pervasive feeling of inquiry that students are aware of, without always being able to explicitly elaborate upon its source. The fact that students envisaged transfer of insights from the articles to the high-school inquiry laboratories also points to advances in inquiry thinking. Documentation of class enactment enabled us to detect episodes when teachers elicited inquiry thinking among their students. For instance, the ability to design experiments and methods suitable for answering research questions is considered a demanding parameter of inquiry thinking (Zohar 2000).

Discipline-Specific Epistemic Beliefs We anticipated that the structure and genre of the research articles would convey the scientists' voice, their conjectures, rationale and enthusiasm, thereby promoting a realistic epistemological perspective of the scientific endeavor in general and of the biotechnologists' perspective in particular. Students' comments during the group interviews emphasized the importance they attributed to the 'authenticity' and the situational aspects of the adapted articles. However, for the students, the article became the research itself, and was regarded as 'real-world' science, rather than a mere reflection of it. The structure of the APL articles led the students to view the nature of science as an ordered process, a faithful example of the 'scientific method'. Moreover, the discussion section raised a sense of discomfort among the students. This is probably because the discussion contradicts the high-school students notion of final conclusions since it reveals, instead, the remaining fluidity and uncertainties of the inquiry process. Nevertheless, hearing the 'scientists' voice' throughout the article provided students with the impression that they were sharing in a professional community with the experts in biology. The article promotes their apprenticeship process by allowing them to become members of this community.

Comprehension of Subject Matter Students and teachers affirmed that the APL genre elicits better comprehension of the content than popular scientific articles, and even textbooks. They linked this benefit to the canonical structure of the APL articles which provides a clear content-organization template, and facilitates the distinction between conjecture, evidence and implications. Such content organization provides students with visible borderlines that appear to stimulate self-regulation of the reading, comprehension, and interpretation processes. Research articles provide students with a naturalistic environment in which relevant but previously acquired biological concepts and ideas are used in a novel context, and prove to be useful for solving problems. Comprehension is boosted by both concretization of the relevant prior knowledge and sometimes abstract ideas, and their application in an authentic situation in a way that emphasizes their relative importance.

Students' Challenges When Learning Biotechnology Through APL Despite the reported positive student and teacher responses, learning through APL is not a trivial task. The main challenge reported was linked to the comprehension of biotechnology processes and methods. Not only were students expected to understand these processes and methods as presented in the articles, but sometimes they were even required to suggest them a priori, via assignments and the conversational model. Complying with these expectations required at a minimum: (i) knowledge integration and application at two levels—prior school biology knowledge integrated and applied in the context of a method and suitable methods to be further integrated into a coherent entity and applied in order to solve a problem; (ii) inquiry thinking—understanding the complex inter-dependency of the research question and the method selected to investigate it and the inter-dependency between the selected method and the results obtained and their interpretation. When students could not grapple with the complexity of the text, they encountered some cognitive overload, and confusion was the result. Many methods in the APL articles are characterized by the fact that they harvest relevant prior biological knowledge, thus leading to a requirement to apply relevant prior knowledge (i.e., knowledge about antibody-antigen complex which is formed in an immune reaction in an organism) in a different context (i.e., antibody-antigen complex formed in the course of protein quantification in the lab).

Interestingly, in an independent study in which a web-based prototype APL was tried in two high-school classes, three of the findings reported above were also identified (Norris et al. 2009): (i) Students' affective engagement; (ii) Discipline-specific epistemic beliefs; and (iii) Comprehension of the subject matter. The web-based APL article was adapted from a PSL article on a mathematical model of the spread of the West Nile virus in North America. The students (n = 62) from two classes participating in the study, were exploring in pairs the web-based APL during an entire 90-min class. They found the biology and mathematics of the West Nile virus to be interesting, similarly to the students who learned using the APL article in biotechnology. Different students reported different interests: the calculus content, the population modeling, and the animal behavior. In addition, the inclusion of a detailed discussion of the assumptions and restrictions of the mathematical model led to some of the most fascinating comments from students, who are not accustomed to being permitted to confront such texts. Many students were surprised to learn that the mathematical model did not mirror the world, but that it was an approximation based upon several compromises, similarly to the promotion of a realistic epistemological perspective of the scientific endeavor observed among the high school students who learned using the APL article in biotechnology. Finally, many students reported that the West Nile virus APL article improved their understanding of topics covered in biology class. Such positive feedback from the students is interesting, since a common complaint by students and teachers is that school mathematics and science classes seem to exist in separate worlds from their daily lives. Many students reported the mathematics remained beyond their grasp, but that they understood the flow of diagrams and sensed a level of satisfaction with that understanding.

In the study performed by Norris et al. (2009), in which the use of a web-based APL in mathematical biology was examined, many students spent considerable time on supplementary problems. These supplementary problems were prepared as peripheral exercises to reinforce the application of basic high school mathematics to a higher level modeling and biological problem solving, but were not crucial to engaging with the conceptual arguments in the prototype article. Nevertheless, many students spent time on these supplementary problems and insufficient time on the more important tasks related to deeper understanding. Thus, learning through APL is not a trivial task whether the topic is mathematical biology or biotechnology. Each discipline has its specialty-specific challenges to learning through APL.

In another study Falk and Yarden (2009) analyzed an 80-min discourse developed during the enactment of an article from the APL-based curriculum in biotechnology in one class (12th grade, religious high school for girls, n = 8), and examined epistemic practices used by the students during their meaning-making of the Results and the Discussion sections of the article. We specifically examined how students connected elements of different epistemic status or context (theory, data, experimental stages, biotechnological applications, and text). Seventy times during the lesson, students were observed connecting elements of different epistemic status or context. Connecting between elements of different epistemic status is not common in science classrooms and it may point to authentic scientific practices that emerge in classrooms that study using APL. We describe briefly the characteristics of the connections that emerged in the course of the analysis of the specific lesson, in which two graphs from the Results section and the Discussion section of an APL article (adapted from Davidov et al. 2000) were discussed.

In-depth analysis of transcripts prepared from video-recording of the lesson and interviews with the teacher revealed that during the lesson students made links between any two of the following four elements: (i) the different stages of the experimental research, as described by the 'scientific method' (the research questions and hypotheses advanced, the research methods used, the conclusions drawn); (ii) the obtained data, as displayed by the graphic representations; (iii) the biological explanations that served as an a-priori theory and assumptions for the research; and (iv) the practical aspects of biotechnology associated with the specific research. The most frequent types of connection between these elements performed in class were:

(a) *Data-theory connection*, causally connecting the data presented in the graphs that appear in the Results section with the biological explanations of these data, namely the theory. The theory is based on students' prior knowledge of the physiology of bacteria, and on the knowledge provided in the Introduction section of the article on the influence of genotoxic materials on bacterial DNA.

(b) *Data-experimental stage connection*, connecting between the data and the experimental stages of the research, i.e., asking research questions, raising hypotheses, designing suitable research methods, performing the experiment, and providing conclusions. While connecting, the students relied on the data presented in the graphic representations in the Results section, on the textual information provided in the Introduction and Methods' sections and on additional information provided by the teacher.

(c) *Data-application connection*, including connections between the data and elements of practical biotechnology, e.g., environmental problems and their possible solutions, practical aims of the research, industrial aspects of the product, and merchandizing the product.

(d) *Experimental stage-application connection*, connecting the experimental stages with the aspects of practical biotechnology.

(e) *Theory-experimental stage connection*, connecting the aforementioned elements of theory with experimental stage and without immediate involvement of the data.

In addition to these types of connections, the analyses revealed that students made links between the text of the article itself, its function, organization, genre and the manner in which it reflects the research process. They connected the text that is being discussed with textual information provided in other sections of the article, or the discussed text with the experimental stage. The most frequent types of connection between these elements performed in class were:

(a) *Text-text* connects the knowledge presented in a specific section or paragraph of the article with knowledge provided in another section. The students used the text that was previously read aloud by one of the students, as well as prior sections of the article discussed in previous lessons.

(b) *Text-research* connects the experimental stages with the way in which they are presented in a specific paragraph of the article.

In the context of the Results section, students mainly made connections between the various elements mentioned above in order to create meaning out of the complexity of the data presented in the graphs, to understand anomalous data points in the graphs, and to make meaning of the unfamiliar methods used for collecting the data presented in the graphs. Interestingly, claims of difficulty or uncertainty made by the students in the context of the meaning-making process of the data and of the experimental and practical elements related to the data could be associated with events in which students were connecting between elements. Since students' claims of difficulty are intended activities that are used by them in the process of acquiring and evaluating knowledge, they may be considered as a type of epistemic practice. This type of authentic practice stems from the students' motivation toward meaning-making of the Results (in contrast to some traditional school situations in which the teacher is the one probing for or even admonishing meaning-making difficulties). In most instances, the claims of difficulty seemed to be based on metacognitive processes, as they emerged from attempts to regulate previously acquired knowledge within the context of the Results. For example, the meaning-making difficulty of one of the graphs could be associated mainly with the data (i.e. anomalous data), the data representation (i.e. two dependent variables plotted on the same graph), the complexity of the research methods by which the data were obtained (the absence of mitomycin during storage in contrast to its presence when assaying the stored bacteria), and the comprehension and interpretation of the research goals.

In the context of the Discussion section, students created meaning for each paragraph by connecting with the information provided in other sections of the article or with the inquiry process performed by the scientists. The students viewed the Discussion as representing a summary of the data presented in the Results section, without being aware of its' argumentative role. The students attempted to assess the novelty of the information provided in the Discussion, by connecting the content with the knowledge they had acquired in other sections of the APL article. Similarly to the abovementioned study in which students who read an APL article in a laboratory setting had difficulty distinguishing between past experiments and the present experiment (Brill et al. 2004), and also in the lesson analyzed in this case, the same difficulty arose while studying the Discussion section.

Taken together, all the above mentioned connections were applied in the lesson. However, the *text-text* and the *text-research* connections were applied almost exclusively in the context of studying the Discussion. It seems that the connections performed by the students in the course of studying the Results and the Discussion sections of the APL article, resemble connections made by scientists when reading scientific texts. Thus, studying through the APL article may enable students to practice authentic scientific practices in schools.

Outcomes of Combining APL with Other Text Genres

In Chap. 2 we outlined the characteristics of various scientific text genres. We specifically focused on four genres, namely Primary Scientific Literature (PSL), APL, Journalistic Reported Versions (JRV), or popular articles published in daily newspapers and science textbooks. We pointed out various differences between these genres, including their organizational structure and the way science is presented in them (see Table 2.1). It is reasonable to assume that the unique characteristics of each text genre may lead to different learning outcomes. In this section, we outline the outcomes of studies in which learning an APL article was compared to or combined with learning using one or more of the other three genres without teacher intervention and in classrooms.

Learning Using Various Scientific Text Genres with No Teacher Intervention

Two scientific text genres are commonly used in high school science lessons: textbooks and popular scientific articles from the daily news (or JRV). The APL genre is relatively new, and therefore not many studies have been conducted in order to compare its use for science learning to the other two commonly used text genres. We describe below two such studies in which the effect of learning using APL and JRV on high school students' comprehension and use of scientific inquiry skills was compared (Baram-Tsabari and Yarden 2005; Norris et al. 2012). The two studies were conducted in two different countries (Israel and Canada), were focused on different disciplinary contexts (high school biology class and high school mathematics class), and had several other differences emanating from the different contexts in which they were carried out, as detailed in Table 7.1 below. Despite these differences, the effect on students' critical thinking was virtually identical in both studies.

The study by Baram-Tsabari and Yarden (2005) used as a platform a breakthrough research article that describes the design of a polyvalent inhibitor to the anthrax toxin (Mourez et al. 2001). This PSL article was translated into Hebrew and adapted to an APL article (an English version of this APL article appears in Chap. 8) and a JRV article. The adaptation process of the two articles preserved most of their content (APL—1,546 words, JRV—828 words). One group of students received the APL article and another group received the JRV article. The way in which the text genre influences the formation of students' scientific literacy was compared between the two groups. Following the reading of one of the two texts, three types of open-ended items were used as assessments tools: (i) Communication of the main ideas and conclusions that were detailed in the article were assessed in the format of a written abstract; (ii) Reading comprehension and acquisition of biological knowledge were tested in the format of content-based True/False questions; (iii) Understanding of the processes and methodology of scientific inquiry were demonstrated in the format of three open-ended questions (dealing with future investigation, critical thinking, and applications of the study). In addition, students' attitudes toward reading APL and JRV were assessed using a Likert-type scale of 11 statements concerning the reading task with a scale from 1 to 6, with 1 being strongly disagree and 6 being strongly agree.

Analysis of 272 questionnaires completed by high school students in Israel who had read one of the two articles revealed that although there was no significant

Table 7.1 A comparison between the experimental setting in two studies that examined the outcomes of learning through APL and JRV

	Baram-Tsabari and Yarden (2005)	Norris et al. (2012)
Location	Israel	Canada
Language of text	Hebrew	English
Article type	Experimental	Theoretical
Class context	Biology	Mathematics
School type	Urban and sub-urban	Urban
Class credit	Yes	No
School grade	10th, 11th, 12th	12th
Number of students	272	211
Gender	70 % females	47 % females

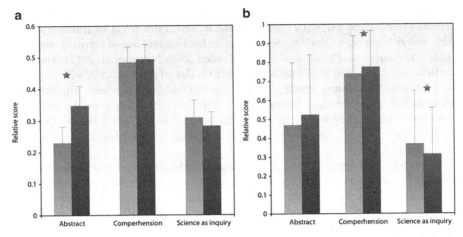

Fig. 7.1 Comparison between learning through APL and JRV articles from the studies of Norris et al. (2012, **a**) and Baram-Tsabari and Yarden (2005, **b**) Students were tested for their ability to write an abstract (Abstract), their reading comprehension (Comperhension) and their inquiry skills (Science as inquiry), after reading an APL article (*Light grey*) or a JRV article (*Dark grey*). Significant differences are marked: *$p < 0.05$. The data was analyzed using a t-test

difference in the students' ability to summarize the main ideas of each text (Abstract in Fig. 7.1, part b), there were significant differences in the students' ability to demonstrate their comprehension and inquiry skills (Comprehension and Science as inquiry, respectively, in Fig. 7.1, part b). Students who read the APL article presented a better understanding of the processes and methodology of scientific inquiry than did those who read the JRV article, while the latter demonstrated better comprehension of the text than did the APL readers. Thus, JRV generated a better understanding of the text, while APL facilitated understanding of how biological research is carried out. Specifically, with regards to understanding the processes and methodology of scientific inquiry; students who read the APL article raised more scientific criticism of the researchers work and methodology and suggested more applications of the technology described in the article than did the students who read the JRV article.

The study by Norris et al. (2012) used as a platform an article that was published by their colleagues (Wonham et al. 2004). This is a theoretical PSL article with an argumentative structure presenting a mathematical model to describe the spread of the West Nile virus. The adaptation to an APL article (4,008 words, see Chap. 9) and a JRV article (3,759 words) was aimed for an introductory differential calculus course offered for 12th grade students in Alberta, Canada. Both articles covered the same main points. The procedure was similar to the one used by Baram-Tsabari and Yarden (2005), namely each student was given either the APL article or the JRV article and they completed a questionnaire. The questionnaire was essentially similar to the one used by Baram-Tsabari and Yarden (2005) apart from the adaptation for the different content.

Analysis of 211 questionnaires completed by 12th grade students in Canada who had read one of the two articles revealed that similarly to the study of Baram-Tsabari and Yarden (2005), the JRV readers better understood the articles while the APL readers thought more critically about the article. In contrast to the study that was carried out in Israel in which students who read the JRV article demonstrated fewer negative attitudes towards it than students who read the APL article, the attitudes of the students in Canada towards reading the articles were positively associated with their comprehension and inquiry skills, regardless of the article they read. Taken together, both studies clearly show that asking students to read text that resembles scientific writing, namely APL, increases their use of critical thinking skills when reading. The fact that the study of Baram-Tsabari and Yarden (2005) was replicated by Norris et al. (2012) in another language, on a different topic, and in a different culture significantly strengthens our claim that reading the APL genre can promote scientific literacy.

In the two studies described above attempts were made to isolate the effect of instruction. Even though the studies were carried out in classrooms, each student was requested to read the text and answer the questions individually and teachers were requested not to be involved in the intervention. This procedure enabled us to conclude that the observed differences between the APL readers and the JRV readers might be the result of the genre differences between the two articles, and not differences emanating from different instruction.

An earlier study by Phillips (1988) identified that background knowledge is useful only in the context of reading strategies. In the context of reading APL, Norris et al. (2012) found that background knowledge, measured by the 12th graders' science courses taken, was not a significant predictor of the score on any of the items used for assessment (i.e., True/False questions). In the study of Baram-Tsabari and Yarden (2005), the best scores were achieved by 12th graders, followed by 11th graders, and then followed by 10th graders, a welcomed pattern. In both studies general knowledge was used in order to examine its possible correlation with students' achievements. However, Yu (2009) reported that familiarity with the general topic of the text (i.e., biology or mathematics) does not facilitate under-standing, but rather knowledge of specific topics contributes to understanding (i.e., exponential growth and decay models; which were taught in advance in the Canadian study). Despite the fact that in the Norris et al. (2012) study students were taught in advance four topics in mathematics that are required in order to understand the articles, it seems that in both studies students' domain specific knowledge was low. In cases where domain knowledge is low, readers rely on other forms of prior knowledge, such as knowledge of text structure to facilitate comprehension. This finding might explain the fact that in both studies readers of the more familiar JRV achieved higher scores on writing the abstract (Fig. 7.1). Thus, knowledge of strategies for dealing with the scientific genre should be taught, since it appears students are not able to acquire such knowledge and strategies without instruction.

In terms of students' attitudes toward the texts, in both studies a strong associ-ation was identified between positive attitude and comprehension and between positive attitude and the science as inquiry scores. These findings are in line with the above mentioned studies of Falk et al. (2008) and Norris et al. (2009) wherein they reported students' affective engagement to be one of the positive outcomes of learning using APL in class, thus providing further support to these findings.

The two studies described in this section took place in schools without any teacher intervention. A more recent study by Braun and Nueckles (2014) compared students who individually read a PSL article (translated into the students' native language, German), an APL article, a JRV article, and a chapter from a science textbook. The study was carried out in a laboratory setting in which high school students (n = 78) were participating alone or in small groups seated at separate desks. Students were asked to read the text and respond to pre- and post-questionnaires. These study findings reveal that students' understanding of the constructive nature of science and the argumentative nature of science significantly improved among students who had read the PSL or the APL articles. The authors of the study concluded that PSL and APL produce more beneficial epistemological reading outcomes than is the case for popular scientific literature and instructional science texts, and made a call to encourage science teachers to incorporate PSL and APL into classroom reading activities.

The outcomes of using APL in classrooms were described in detail in the previous section. In the subsequent section, we describe attempts to combine APL with other text genres, and to compare the use of APL to the use of the other text genres in class.

Learning Using Various Scientific Text Genres in Class

Each of the APL-based curricula that were developed by us (Falk et al. 2003; Yarden and Brill 1999) includes an introduction designed to expose the high school students to the basic concepts and processes of each field in addition to three to four APL articles (Yarden et al. 2001). The introduction to each curricula is written in a textbook genre, thus in classes that study using the APL-based curricula students learn the topic, either developmental biology or biotechnology, using two different genres: the textbook genre and the APL genre. Such classes enable the examination of possible differences between learning using each of the genres in the same class context.

In a study conducted by Brill and Yarden (2003), students who studied developmental biology using the APL-based curriculum were asked before starting to study the topic, after learning the introduction to the curriculum, and following the completion of one APL article from the curriculum, what they found interesting to know about embryonic development. In addition, questions asked orally by the students during the lessons were also collected and analyzed. The questions were classified into three categories (following Dillon 1984): properties, comparisons, and causal relationships. These three categories were previously suggested by Dillon (1984) as first-, second-, and third-order categories, respectively, and indicate their relative increase in contributing to scientific knowledge. The causal relationship questions are considered the highest order of research questions, since answering questions in this category requires finding the relation, correlation, conditionality, or causality of the subjects in question. In most cases, an experiment is needed in order to answer questions in this category.

Analysis of the questions collected in the course of this study revealed that before learning students tended to ask only questions of the properties category. However, after reading one APL article, the students tended to pose more causal relationship type questions. In addition to the observation that high order level questions were not often asked by students when learning the introduction to the curriculum, such questions were rarely detected during and following instruction using a textbook in genetics, which took place during the same time frame. The collected questions were also coded according to content and specifically to either similar or identical and unique types of questions. In addition to the increase in questions of the third-order category, which occurred following learning an APL article, an increase in the number of unique questions was also observed. In contrast, a decrease in the number of unique questions was observed in the class that studied genetics using a textbook. Taken together, high school students tend to pose questions that reveal a higher level of thinking and uniqueness during or following instruction with APL. This change was not observed during or following instruction with a textbook. We therefore suggest that learning through APL may be one way to provide a stimulus for question-asking by high-school students that may result in higher thinking levels and uniqueness, while learning using a traditional textbook genre did not similarly cause such a stimulus. Since a scientific research paper poses a research question, demonstrates the events that lead to the answer, and poses new questions, we suggest that learning through APL may be one way to develop the basic skill of question-asking among high school students.

All the studies described so far were carried out among high school students, mostly in the 12th grade, and as low as in the 10th grade. Shanahan (2012; Shanahan et al. 2009) attempted to extend the use of APL beyond the high-school science classroom to the elementary school level. Toward this end, they used texts that integrate APL with narrative writing about science and scientists and termed it Hybrid Adapted Primary Literature (HAPL). In contrast to the APL genre, the narrative part is not written in a passive voice, it does not include abstract noun phrases, and it directly identifies the people involved and describes their thoughts, motivations, and actions. It is aimed to provide the elementary school students with a more complete picture of the nature of science than that provided by APL. A version of HAPL in geology was examined with a single 6th grade class in Canada. It was found that the HAPL engaged the elementary school students in reading and that it helped to develop students' understanding of scientists and scientific inquiry. The students also recognized through reading the HAPL that there is much difficult and time-consuming work and effort by scientists in the doing of their research. Students were also reported to recognize the dynamic nature of scientific knowledge (Shanahan et al. 2009).

An aspect of attempting to use APL at the elementary school level includes attempts to use this text genre at the middle school level. In curriculum that was developed for 8th grade students, three short APL articles with increasing complexity were developed (Ariely and Yarden 2013) and they are currently under examination in schools in Israel. There are translated versions of these APL articles and they are currently under examination in the US (Davis and Demir 2015). The APL articles that are included in this 8th grade curriculum are short (approximately 1,000 words in length), which were adapted to the knowledge level of 8th graders, and they are accompanied with a JRV article. A set of activities lead the students to compare the differences between the two text genres, thus allowing them to recognize the authentic scientific genre as part of the learning process. These studies are expected to provide the basis for developing a learning progression for the use of various scientific text genres in science classrooms from the elementary school level, through middle school and on into the high school level.

Suggested Learning Progression for the Use of APL

One of the eight practices mentioned in the framework for K-12 science education (National Research Council [NRC] 2012) includes "Obtaining, evaluating, and communicating information." In order to become proficient in this practice 12th grade students are required, among other requirements, to "engage in a critical reading of primary scientific literature (adapted for classroom use) or of media reports of science and discuss the validity and reliability of the data, hypotheses, and conclusions" (p. 76). The description of the progression towards this goal in the framework mentions, that in high school only these practices should be further developed by providing students with more complex texts like appropriate samples of APL that may enable them to begin seeing how science is communicated by science practitioners.

The described studies that examined the use of APL, alone or in conjunction with other text genres, suggest that APL might be used for science learning earlier than at the high school level. We suggest that as long as the content of the text is adapted to the cognitive level and relevant prior knowledge of the target population, APL can be used for science learning also at the upper levels of elementary school (5th and 6th grades), as well as in middle school. The experience of Shanahan et al. (2009) with 6th grade students who learned using HAPL, as well as through their own experience in using short APL articles along with short JRV articles with 8th grade students, allow us to suggest that short and simple APL articles can be used at an earlier stage in the learning process. HAPL allows the students to become familiar with the structure of scientific writing early in their education and prepares them to read more complex texts later in their education. This suggestion is also in line with one of the key requirements of the Common Core State Standards for Literacy in Science and Technical Subjects (Achieve 2012), namely that when reading scientific and technical texts, students need to be able to gain knowledge from challenging texts that often make extensive use of elaborate diagrams and data to convey information and illustrate concepts. Reading Standard 10 asks students to read complex informational texts in these fields with independence and confidence.

References

Achieve. (2012). *Next generation science standards.* Available: http://www. nextgenerationscience.org. October 28 2014.

Alexander, P. A., & Jetton, T. L. (2000). Learning from text: A multidimensional and developmental perspective. In M. L. Kamil, P. B. Mosenthal, P. D. Pearson, & R. Barr (Eds.), *Handbook of reading research* (Vol. 3, pp. 285–310). London: Lawrence Erlbaum Assoc. Pub.

Ariely, M., & Yarden, A. (2013). Exploring reproductive systems. In B. Eylon, A. Yarden, & Z. Scherz (Eds.), *Exploring life systems (Grade 8)* (Vol. 2). Rehovot: Department of Science Teaching, Weizmann Institute of Science.

Asher, S. R. (1980). Topic interest and children's reading comprehension. In R. J. Spiro, B. R. Bruce, & W. F. Brewer (Eds.), *Theoretical issues in reading comprehension* (pp. 525–534). Hillsdale: Lawrence Erlbaum Associates.

Ausubel, D. P. (1963). *The psychology of meaningful verbal learning.* New York/London: Grune and Stratton.

Baram-Tsabari, A., & Yarden, A. (2005). Text genre as a factor in the formation of scientific literacy. *Journal of Research in Science Teaching, 42*(4), 403–428.

Braun, I., & Nueckles, M. (2014). Scholarly holds lead over popular and instructional: Text type influences epistemological reading outcomes. *Science Education, 98*(5), 867–904.

Brill, G., & Yarden, A. (2003). Learning biology through research papers: A stimulus for question-asking by high-school students. *Cell Biology Education, 2*(4), 266–274.

Brill, G., Falk, H., & Yarden, A. (2004). The learning processes of two high-school biology students when reading primary literature. *International Journal of Science Education, 26*(4), 497–512.

Davidov, Y., Rosen, R., Smulsky, D. R., Van Dyk, T. K., Vollmer, A. C., Elsemore, D. A., LaRossa, R. A., & Belkin, S. (2000). Improved bacterial SOS promoter: Lux fusions for genotoxicity detection. *Mutation Research, 466,* 97–107.

Davis, M. B., & Demir, K. (2015). *How a teacher uses reading combined text genres to influence meaning construction in science.* Annual conference of the National Association of Research in Science Teaching (NARST). Chicago, USA.

Dillon, J. T. (1984). The classification of research questions. *Review of Educational Research, 54* (3), 327–361.

Epstein, H. T. (1970). *A strategy for education.* Oxford: Oxford University Press, Inc.

Falk, H. (2009). *Characterizing and scaffolding the enactment of adapted primary literature based high-school biology curricula.* PhD thesis, Feinberg Graduate School, Weizmann Institute of Science, Rehovot, Israel.

Falk, H., & Yarden, A. (2009). "Here the scientists explain what I said." Coordination practices elicited during the enactment of the results and discussion sections of adapted primary literature. *Research in Science Education, 39*(3), 349–383.

Falk, H., Piontkevitz, Y., Brill, G., Baram, A., & Yarden, A. (2003). *Gene tamers: Studying biotechnology through research* (In Hebrew and Arabic, 1st ed.). Rehovot: The Amos de-Shalit Center for Science Teaching.

Falk, H., Brill, G., & Yarden, A. (2008). Teaching a biotechnology curriculum based on adapted primary literature. *International Journal of Science Education, 30*(14), 1841–1866.

Holliday, W. G., Yore, L. D., & Alvermann, D. E. (1994). The reading-science learning-writing connection: Breakthroughs, barriers and promises. *Journal of Research in Science Teaching, 31*(9), 877–893.

Kintsch, W. (1989). Leaning from text. In L. B. Resnick (Ed.), *Knowing, learning, and instruction* (pp. 25–46). Hillsdale: Lawrence Erlbaum.

Mourez, M., Kane, R., Mogridge, J., Metallo, S., Deschatelets, P., Sellman, B., Whitesides, G., & Collier, R. (2001). Designing a polyvalent inhibitor of anthrax toxin. *Nature Biotechnology, 19*, 958–961.

National Research Council [NRC]. (2012). *A framework for K-12 science education: Practices, crosscutting concepts, and core ideas*. Washington, DC: The National Academies Press.

Norris, S. P., Macnab, J. S., Wonham, M., & de Vries, G. (2009). West Nile virus: Using adapted primary literature in mathematical biology to teach scientific and mathematical reasoning in high school. *Research in Science Education, 39*(3), 321–329.

Norris, S. P., Stelnicki, N., & de Vries, G. (2012). Teaching mathematical biology in high school using adapted primary literature. *Research in Science Education, 42*(4), 633–649.

OECD. (2012). *The PISA 2015 assessment framework: Key competencies in reading, mathematics and science*. http://www.oecd.org/pisa/pisaproducts/pisa2015draftframeworks.htm

Phillips, L. M. (1988). Young readers' inference strategies in reading comprehension. *Cognition and Instruction, 5*(3), 193–222.

Samuels, S. J. (1983). A cognitive approach to factors influencing reading comprehension. *Journal of Educational Research, 76*, 261–266.

Shanahan, M. C. (2012). Reading for evidence in hybrid adapted primary literature. In S. P. Norris (Ed.), *Reading for evidence and interpreting visualizations in mathematics and science education* (pp. 41–63). Rotterdam: Sense Publishers.

Shanahan, M. C., Santos, J. S. D., & Morrow, R. (2009). Hybrid adapted primary literature: A strategy to support elementary students in reading about scientific inquiry. *Alberta Science Education Journal, 40*(1), 20–26.

Sweller, J., & Chandler, P. (1994). Why some material is difficult to learn. *Cognition and Instruction, 12*(3), 185–233.

Tamir, P. (1985). Content analysis focusing on inquiry. *Journal of Curriculum Studies, 17*(1), 87–94.

Wonham, M. J., de Camino-Beck, T., & Lewis, M. A. (2004). An epidemiological model for West Nile virus: Invasion analysis and control applications. *Proceedings of the Royal Society of London, Series B: Biological Sciences, 271*(1538), 501–507.

Yarden, A., & Brill, G. (1999). *The secrets of embryonic development: Study through research* (In Hebrew and Arabic, 2004 4th ed.). Rehovot: The Amos de-Shalit Center for Science Teaching.

Yarden, A., Brill, G., & Falk, H. (2001). Primary literature as a basis for a high-school biology curriculum. *Journal of Biological Education, 35*(4), 190–195.

Yu, G. (2009). The shifting sands in the effects of source text summarizability on summary writing. *Assessing Writing, 14*(2), 116–137.

Zohar, A. (2000). Inquiry learning as higher order thinking: Overcoming cognitive obstacles. In J. Minstrell & E. H. van Zee (Eds.), *Inquiry into inquiry learning and teaching in science* (pp. 405–424). Washington, DC: American Association for the Advancement of Science.

Part III
A Compendium of Adapted Primary Literature Annotated for Critical Reading

Chapter 8
Developing an Inhibitor of Anthrax Toxin: Annotated Examples of Adapted Primary Literature (APL) and Journalistic Reported Version (JRV) Articles

In this chapter we provide two annotated examples that originated from a single Primary Scientific Literature (PSL) article, reporting about the design and testing of an inhibitor of Anthrax toxin (Mourez et al. 2001). The PSL article does not include a specific research question, rather it describes a technological achievement along with its biological basis. First, it reports on the identification of a peptide that binds the toxin and weakly inhibits its enzymatic action. Subsequently, copies of this peptide were linked to a backbone and the resulting polyvalent molecule was shown to strongly inhibit the toxin action. The PSL article has the following organizational structure (from the beginning of the article to its end): Abstract (without a header), Introduction (without a header), Results, Discussion, and Experimental protocol.

Two articles, APL and JRV, were developed from this PSL article and they appear in respective order. These two articles were used in the research study of Baram-Tsabari and Yarden (2005) described in detail in Chap. 7. A comparison between the two articles enables us to point out the unique features of each of them and especially highlight the unique features of APL (see annotations that accompany this chapter).

The APL article, which appears first in this chapter, includes the classical canonical organizational structure of PSL, namely Introduction, Methods, Results, and Discussion. This article lacks an Abstract, since in the course of the research that was carried out using these two articles the participating students were asked to communicate the main ideas that were detailed in the articles in a format of a written abstract (see Chap. 7). The glossary that appears at the end of the APL article provides explanations for scientific terms that appear in a bold face along the text. These scientific terms were identified as terms that might not be familiar to high-school students, and the explanations appeared at the margins of the text when handed to the students and not at the end of the article. The endnotes that are provided at the end of the APL article highlight features that are discussed in detail in Chaps 2, 3, and 4 and especially those related to epistemology and meta-scientific language.

© Springer Science+Business Media Dordrecht 2015
A. Yarden et al., *Adapted Primary Literature*, Innovations in Science
Education and Technology 22, DOI 10.1007/978-94-017-9759-7_8

In contrast to the PSL and APL articles, the JRV article lacks an organizational structure, but it does include the table and two of the three figures that were included in the APL. Such illustrations are not always included in JRV articles. Apart from the absence of an organizational structure that can be easily identified in the APL article, the JRV article is written in a popular manner, that is it lacks a glossary and includes information about the scientists that conducted the research study, as pointed out in the annotations below.

Developing an Inhibitor of Anthrax Toxin[1]

Ayelet Baram-Tsabari[*] and Anat Yarden
Department of Science Teaching, Weizmann Institute of Science, Rehovot, Israel

Introduction

Anthrax **toxin** is produced by the bacterium *Bacillus anthracis* and is responsible for most of the **symptoms** of anthrax disease. Anthrax is a rare disease in humans, infecting those who come in touch with cattle, ships or goats and their products (meat, wool, and skins). Recently, concerns of using *Bacillus anthracis* in weapons and biological terrorism were confirmed.

> **Goal-Directed Structure**
> In the first paragraph of the APL article, the Anthrax toxin and the bacterium that produces it are introduced as well as with the justification for carrying out the reported study. The aim of the last sentence in this paragraph ("Recently, concerns of using *Bacillus anthracis* in weapons and biological terrorism were confirmed") is to make a case for the relevance of the research.

A few characteristics of the anthrax bacteria make them suitable for use in biological warfare: they can be spread through aerosol (a suspension of microscopic droplets that contain the bacteria); the number of disease-causing bacteria is

---

[1] This APL article is an adaptation based upon Mourez et al. (2001).

[*]This adaptation was completed when Ayelet Baram-Tsabari was a PhD student at the Weizmann Institute of Science.

comparatively small (8,000–10,000 bacteria); the resulting disease carries a high death rate; the bacteria can transform into spores, resistant to being carried in a missile and easy to be spread by the wind; last, the bacteria are easily found in soil. On the other hand, anthrax is non-infectious from human to human.

Argumentative Structure
The second paragraph of the APL article presents a theoretical argument supported with several characteristics of experimental evidence as to why Anthrax is suitable as a biological weapon.

As the spores invade the body they are captured by **phagocytes** of the immune system. The spores then germinate (transform into active bacteria) in lymph nodes or in close proximity to the invaded organs. The living bacterium produces toxins that cause damage to surrounding tissues, edema and hemorrhages, due to changes in capillary permeability. The disease has three forms in humans, depending on the organ invaded by the spores: skin anthrax, intestinal anthrax and respiratory anthrax. An anti-anthrax vaccine was developed already in the 1960s but it is not practical to vaccinate the entire population. Antibiotic treatment can eradicate the bacteria in an infected person, but cannot affect the toxin that has been produced. Thus, developing a specific inhibitor to the toxin might yield a beneficial addition to the use of antibiotics.

Argumentative Structure
The third paragraph describes the infection process and presents an additional argument to support the development of an inhibitor to the toxin. The argument is based on experimental evidence that antibiotic treatment of the bacterial infection does not prevent the production of the toxin in the course of the infection.

Epistemology (Uncertainty)
In the last sentence in the third paragraph the word "might" is used when referring to the possible benefits of developing the specific inhibitor to the Anthrax toxin. The use of the word "might" implies that the scientists are tentative and uncertain of the information given, namely the development of the inhibitor might be beneficial, but it can also prove to be non-beneficial. This example is in line with Chap. 2 in which we discussed the uncertainty as one characteristic of scientific text genres, such as PSL and APL. Other terms used in the literature for this text feature include modality, hedging, or tentativeness.

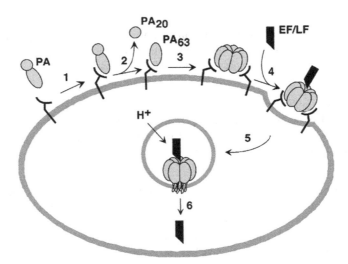

Fig. 8.1 Assembly of the anthrax toxin and its entry into cells. (*1*) The PA protein binds a **receptor** on the cell membrane. (*2*) A portion of the PA protein is cleaved. (*3*) Seven units of the cleaved PA protein join together to form a channel in the cell membrane. (*4*) The EF and LF enzymes bind the channel. (*5*) Endocytosis mediates the entrance of the protein complex into the cell. (*6*) The death- causing enzymes are released to the cytoplasm

The anthrax toxin is composed of three proteins: PA is responsible for the attachment of the toxin to the cell membrane and its cellular invasion. EF and LF are enzymes that cause cell death. The bacteria secrete these three proteins separately, in their non-toxic form. By diffusion they reach the patient's cell membrane and assemble to form the active toxin (Fig. 8.1).

The assembly is enabled through an additional enzyme (**protease**) that is located on the cell membrane. This enzyme cleaves a portion of the PA protein; seven cleaved PA proteins assemble to create a heptamer, which strongly adheres to the EF and LF enzymes. The resulting protein complex enters the cell through endocytosis and is trafficked to the **endosome**. In the acidic environment of the endosome, the PA protein enables the two enzymes to enter the cell cytoplasm. Both enzymes are not active by themselves, but are made active when attached to the PA protein: PA + EF cause tissue edema while PA + LF lead to the inactivation of an important system responsible for cellular growth and development. In this article we will focus on the PA + LF couplet, ignoring the role of the EF protein.

This article will describe the development and testing of an inhibitor of anthrax toxin that works by binding to the PA protein **heptamer** and preventing it's binding to the LF enzyme. A peptide (a short protein) that can bind the cleaved PA protein, and thus could weakly prevent its binding to the LF enzyme, was chosen out of a large pool of anti-bacterial viral agents (**phages**). Assembling multiple copies of this peptide to form a single molecule produced a protein construct that could efficiently inhibit toxin formation, and was thus proved to inhibit the assembly and activity of the toxin in **cell cultures** and in rats.

Meta-scientific Language

The last paragraph of the Introduction outlines the aim of the study, namely "the development and testing of an inhibitor to anthrax toxin that works by binding to the PA protein heptamer and preventing its' binding to the LF enzyme." This paragraph describes the research in a nutshell and sets the stage for the entire study. In the last sentence the word proved is used when referring to multiple copies of a peptide that were isolated and that these multiple copies proved to inhibit the assembly and activity of the toxin in cell cultures and in rats. Providing proof for or evidence to support the hypothesis is at the basis of scientific research and therefore it is frequently used in PSL and APL. Accordingly, multiple meta-scientific language terms can be identified in this APL article and this is just one of them.

Materials and Methods

Choosing the Peptide One method for inhibiting the action of the anthrax toxin is by interfering with the assembly of its constituents. The search started with finding a phage that binds to the cleaved and activated PA protein in the vicinity of the site where the PA naturally binds with the LF enzyme so it could interfere with their interaction, while excluding proteins that bind to the uncleaved PA protein (Fig. 8.2).

A cleaved PA protein was anchored to a plastic surface; phages that express a large variety of peptides were then added. After an incubation period, the surface was washed in order to discard phages that do not express peptides that bind to the PA protein. Next, an uncleaved PA protein was added to the surface, binding peptides that adhere to it better than to the cleaved form; these, too, were washed out following an incubation period. Phages that were eventually left on the surface were those that express peptides that bind the cleaved PA protein but not the uncleaved.

Designing the Inhibitor A molecule composed of multiple copies of the peptide, chosen as described above, was formed and assembled on a backbone of a flexible synthetic material. Each molecule carried an average of 22 such peptides.

Assessing the Inhibiting Activity in Cell Culture This assessment was carried out by adding the inhibitor to the growth-media of a culture of hamster ovary cells along with the constituent proteins of anthrax toxin. In order to confirm molecular binding, molecules were tagged with radioactive markers. In order to estimate the ability of the inhibitor to inhibit the biological activity of anthrax toxin, the ovarian cells were incubated with PA and LF proteins that carry a portion of the diphtheria toxin. This method allows the researchers to learn when successful **endocytosis**

Fig. 8.2 Choosing the peptide. (*1*) A heptamer of the cleaved PA protein was anchored to the plastic surface. (*2*) Phages that express a variety of peptides were added to the solution. Phages that did not express a PA-binding peptide were washed out. (*3*) Uncleaved PA proteins were added to the solution to identify and discard peptides that bind the uncleaved form as well. (*4*) The remaining phages were collected by a non surface-anchored heptamers

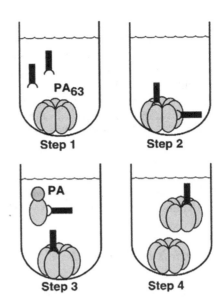

occurs and the toxin invades the cells. If it invades, the diphtheria toxin stops the cellular protein production, which can be easily demonstrated in the lab.

Assessing the Inhibiting Activity in Animal Models Finally, proving the therapeutic potential of the inhibitor was performed by injecting the anthrax-toxin into rats. A certain rat specie to which the combination of the PA and LF proteins is known to be highly lethal, leading to death within a few hours, was used. The ability of the inhibitor in preventing the onset of symptoms caused by the toxin was assessed.

Epistemology (Passive Voice)
The Materials and Methods section is written in a passive voice. For example, it is written that "A cleaved PA protein was anchored to a plastic surface..." Thus illustrating that the authors intentionally did not write who carried out the specific action that led to the anchoring of the PA protein to the plastic surface. Use of the passive voice is common in PSL articles and especially in the Methods and the Results sections because it affords presenting the information as objectively as possible by distancing it from the authors as agents. In addition, it helps to generalize the methodology used in the study, since the Materials and Methods section provides information that may enable other scientists to repeat the experiments elsewhere.

Results

Isolation of an Inhibiting Peptide and Assessment of Its Activity In order to inhibit the activity of anthrax toxin, a protein that interferes with the assembly of its constituents was sought; a protein that binds the PA protein and will not allow further binding to the LF enzyme. It was thus important that the protein will bind only the cleaved and activated form of the PA protein, in the vicinity of the binding site of both enzymes. After repeating the searching process three times, two peptides that bind to the cleaved but not to the uncleaved PA protein heptamer were found. When the LF enzyme was added to the solution, the binding of these two peptides was inhibited. This observation supports the hypothesis that the peptides bind the PA protein at the same site as the LF enzyme does, or at least in the vicinity.

Meta-scientific Language
The last sentence in the first paragraph of the Results section includes two meta-scientific terms: "observation" and "hypothesis". The use of these terms enables the authors to connect their observation to the Introduction section where the complex structure of the Anthrax toxin is described in detail.

Further analysis revealed that both peptides contain an identical sequence of four **hydrophobic amino acids**. It is thus reasonable to assume that these amino acids are responsible for binding the heptamer, despite the fact that this sequence of four amino acids is not found in the LF enzymes.

Epistemology (Uncertainty)
Even though the authors showed that "both peptides contain an identical sequence...." They chose to present this finding as tentative, or uncertain, by writing "It is thus reasonable to assume that these amino acids are responsible for binding the heptamer..." The reservation the authors chose to express through use of the words "reasonable to assume" provides another example of the epistemology of scientific writing that prefers to present the information as uncertain, thus allowing other scientists to evaluate the information and decide whether the evidence is strong enough to support the authors' assumptions.

Copies of one of the peptides were synthesized and added to the culture media of hamster ovary cells. The peptide was found to have a mild interfering effect with the assembly of the LF enzyme and the cleaved PA protein. A control peptide, under the same conditions, did not interfere with the assembly.

Fig. 8.3 Inhibition of toxin action in cell culture. The effect of various amounts of the single peptide (□), the inhibitor (●) and the control of the flexible backbone alone (■). (**a**) Inhibition of the LF enzyme association with the PA protein. Binding was estimated using radio-labeled protein tags. (**b**) Inhibition of toxin toxicity in cell culture

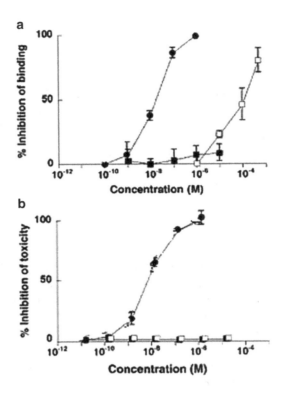

Developing an Effective Inhibitor An inhibitor composed of 22 molecules of the inhibiting peptide carried on a flexible backbone was produced. The inhibitor was added to the growth media of a culture of hamster ovary cells (Fig. 8.3a).

The inhibitor was found to interfere with the assembly of the LF enzyme with the PA protein more efficiently than the single peptide and thus at lower concentrations. In fact, its inhibition was 7,500 times stronger than that of the single peptide. The control substance, the flexible backbone without the peptides, showed no inhibition of the assembly.

Preventing Toxicity in Cell Culture The next stage was to prove that the inhibitor effectively inhibits the biological activity of the toxin. Hamster ovary cells were cultured with the addition of PA and LF protein fused with a portion of diphtheria toxin protein. The diphtheria toxin part is used to show whether the proteins are able to enter the cells or not; if inside the cell, it paralyzes cytoplasmic protein synthesis. Adding the inhibitor protein to the growth media prevented the activity of the diphtheria toxin, while the control substance (the naked flexible backbone) or the single peptide did not affect its toxicity (Fig. 8.3b).

Table 8.1 Inhibition of anthrax toxin action in animal model

Inhibitor	Amounts (in nanomoles)			Outcome
	PA	LF	Peptide	
None	0.5	0.1	–	Symptoms of intoxication
Peptide + backbone	0.5	0.1	75	Symptoms of intoxication
Inhibitor	0.5	0.1	12	Delayed intoxication
Inhibitor	0.5	0.1	75	No symptoms

Table 8.1. All rats were anesthetized and injected with similar doses of the LF + PA solution, along with an additional substance depending on their assigned group (each group consisted of four rats): (1) No inhibitor: the rats showed symptoms of intoxication 1 h after injection. (2) Free peptide + free backbone (not assembled): symptoms were shown 1 h after injection. (3) Small doses of the inhibitor: symptoms were delayed and shown 2 h after injection. (4) High doses of the inhibitor: no symptoms were shown over a 1 week follow-up

Meta-scientific Language
This paragraph includes several meta-scientific terms namely, prove, effectively, and affect.

Preventing the Activity of the Toxin in Animal Models The last experiment, aimed at proving the therapeutic potential of the inhibitor, was performed on live rats, injected with a high dose of the PA and LF proteins, which are responsible together for the lethal symptoms of anthrax (Table 8.1). Injection of the inhibitor along with the toxin proteins effectively delayed the onset of symptoms in rats. Escalating doses of the inhibitor could eliminate the symptoms completely. The rats were also protected when the inhibitor was injected 3–4 min after the toxin proteins. The control substance and the single peptide did not affect the symptoms produced in the rats by the toxin. A 1 week follow-up of the rats did not reveal toxic side-effects for the inhibitor itself.

Discussion

Anthrax toxin, produced by the bacteria *Bacillus anthracis*, causes most of the symptoms of anthrax disease. The toxin is composed of three proteins that assemble on the cell membrane to form the active molecule. Aiming to interfere with the activity of anthrax toxin that remains active in the patient's body even after eradicating the producing bacteria with antibiotics, a peptide that binds to one of its proteins (PA) was located. This peptide prevents the association of the PA protein with the toxin LF enzyme and therefore prevents the formation of the active toxin.

The peptide was then used in order to produce a more efficient inhibitor; its efficacy was confirmed both in cell culture and animal models. In cell cultures, it successfully prevented the assembly of the toxin subunits and its entrance to the cells. While injecting the toxin subunits to sensitive rats, co-injection of the inhibitor in small doses (amounts) could delay the onset of symptoms and even completely eliminate them in high doses.

The inhibition of intoxication observed in this study is the result of specific binding of the inhibitor to the cleaved PA protein on the membrane of living cells and not a result of association between the inhibitor and the PA protein or LF enzyme in the solution. The cleavage of the PA protein and the assembly of the seven resulting products were not affected by the inhibitor (data not shown).

The efficacy of the inhibitor in preventing the activity of the anthrax toxin in animal models suggests that it, or another inhibitor developed by a similar approach, may serve as a therapeutic agent for patients infected with anthrax.

Epistemology (Uncertainty)

The last paragraph of the article presents the findings of the study as tentative. The authors claim that "The efficacy of the inhibitor in preventing the activity of the anthrax toxin in animal models suggests that...it may serve as a therapeutic agent for patients with anthrax." The use of the words, may serve, suggests and may in this sentence demonstrates how uncertainty is expressed in the article, as commonly appears in scientific articles.

Glossary

Cell culture a laboratory technique of growing a population of cells under controlled conditions of growth medium nutrients and oxygen.

Endocytosis internalization of molecules into the cell by coating them in a portion of the cell membrane that folds inside.

Endosome a cellular compartment with an acidified environment.

Heptamer a structure composed of seven sub-units.

Hydrophobic amino acids amino acids that "dislike" water environment and "like" to hide inside a protein.

Phages viruses that can infect bacteria and multiply within them.

Phagocyte a white blood cell of the immune system that is capable of scavenging and digesting bacteria.

Protease an enzyme capable of breaking down the peptide bonds between amino acids.

Receptor a cellular molecule that is responsible for signal reception. In this case the receptor is anchored to the outer cell membrane.

Symptom a sign sense or finding that indicates a certain disease.

Toxin a poisonous substance excreted by living organisms such as bacteria.

A Successful Experiment on the Way to Find a Drug for Anthrax[2]

Ayelet Baram-Tsabari[*] and Anat Yarden

Department of Science Teaching, Weizmann Institute of Science, Rehovot, Israel

Epistemology

The title of the JRV article, namely "A successful experiment on the way to find a drug for anthrax", is aimed to attract the readers of the article to continue reading the text and therefore it presents the scientific information as certain, in contrast to the uncertain way this information is presented in the PSL and APL articles. Note the pronounced difference between the title of the JRV article and the title of the APL article: "Developing an inhibitor of anthrax toxin". This latter title does not claim any success in finding the inhibitor, it claims only the process of developing an inhibitor is underway.

There is no longer a need to explain why finding a drug to anthrax is so important. Even though naturally, the disease is rare among humans and infects mainly those who come in touch with cattle, ships or goats and their products, the use of the bacteria in weapons and biological terrorism makes it an actual public threat. Antibiotic treatment for anthrax is effective only if administered at the early stages of the disease, before the causing bacteria, *Bacillus anthracis*, excretes a sufficient amount of its toxin. A toxin is a poisonous substance produced by a living organism. In the case of anthrax disease, it is also responsible for most of the symptoms, and therefore it is crucial to find a drug that can neutralize it.

Scientific Language

The first paragraph of the JRV article demonstrates clearly that this article is aimed at the general public and was written by journalists and not by scientists, as the meaning of the term JRV hints.: It is a popularization written by science journalists. In contrast to the APL article in this chapter, the JRV

(continued)

---

[2] This JRV article (or popular scientific article) was written based upon Mourez et al. (2001).

[*] This article was completed when Ayelet Baram-Tsabari was a PhD student at the Weizmann Institute of Science.

article does not include scientific terms deemed unfamiliar to the general public, and if included, they are explained immediately. For example, the term toxin is explained as it appears for the first time "A toxin is a poisonous substance produced by a living organism." This is in contrast to the APL article in which scientific terms appear in the text, similarly to the way they appear in a PSL article which is aimed at scientists who are familiar with the scientific terms. However, since the APL is aimed at high school students, explanations for the terms appear beside the text, or at the end of the text in a glossary.

In the past recent months, a group of scientists from Harvard University has achieved an important progress by developing a drug that inhibits the assembly of the subunits of the toxin and proving that it prevents intoxication in rats. The toxin that the anthrax bacteria produce is composed of three proteins: the PA, which is responsible for the attachment of the toxin to the cell membrane and its entrance to the cell, and the two lethal enzymes EF and LF. The bacteria excrete each of these proteins separately, in their non-toxic forms, which diffuse to the patient's cell membranes where they form together the active toxin (Fig. 8.4).

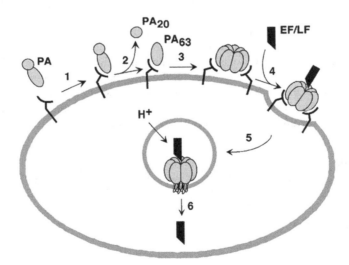

Fig. 8.4 Assembly of the anthrax toxin and its entry into cells. (*1*) The PA protein binds a receptor on the cell membrane. (*2*) A portion of the PA protein is cleaved. (*3*) Seven units of the cleaved PA protein join together to form a channel in the cell membrane. (*4*) The EF and LF enzymes bind the channel. (*5*) The complex enters the cell coated by a portion of the cell membrane. (*6*) The lethal enzymes are released to the cytoplasm

Epistemology
In contrast to the PSL and APL articles which attempt to present the scientific information as objective as possible (for example by the use of passive voice and elimination of human participants), the JRV article overtly presents the scientists who conducted the study in the first sentence of the second paragraph: "a group of scientists from Harvard University has achieved an important progress..." This sentence puts the scientists who conducted the research at the center of the scientific achievement, in contrast to the way they are presented in a PSL or APL articles. Moreover, this paragraph demonstrates another contrast between APL and JRV in terms of the presentation of the information. In the APL article the findings are presented as uncertain, in the JRV the authors present no doubt that in the course of the research an inhibitor to Anthrax toxin was identified.

The assembly of the toxin is enabled through an additional enzyme (a protease, an enzyme that cleaves proteins) that is located on the cell membrane. The enzyme cleaves a portion of the PA protein; seven cleaved PA proteins assemble together to form a heptamer, which strongly adheres to the EF and LF enzymes. This large complex enters the cell through endocytosis—a portion from the cell membrane folds inwards and leaves the membrane, carrying the toxin with it into the cell. Moving in a vesicle, once inside the cell, the PA heptamer releases the two lethal enzymes. Thus, the PA protein enables the two enzymes to invade the cytoplasm where they can act: PA + EF cause tissue edema while PA + LF lead to the inactivation of an important system responsible for cellular growth and development. For simplicity, from here on we focus on the PA + LF couplet, ignoring the role of the EF protein. Attempting to prevent the lethal activity of the anthrax toxin, the researchers decided to try to interrupt with the assembly of its subunits. The idea was to find a protein that binds the cleaved and active PA protein, close to the binding site of the LF enzymes, where it can interrupt with their binding.

The researchers found two peptides (short proteins) that bind the PA heptamer. When they added the lethal portion of the LF protein to the solution, it interrupted with the binding of the peptides to the heptamer. Hence, the researchers concluded that the peptides bind the PA protein at the same site where the toxic enzymes are supposed to be bound, or at least in its vicinity.

In order to find out whether the peptide effactually inhibits the assembly of the toxin in living cells, the researchers added the heptamer, the LF protein and the inhibiting peptide to hamster ovary cells grown in culture. They found out that the peptide conveyed a mild inhibiting activity to the binding of the LF enzyme and the cleaved PA protein; adding a different peptide that served as a control, did not yield a similar effect.

Still, it was clear that the mild interference with the formation of the toxin is not good enough to save from the deadly effect of anthrax. An efficient drug should have a much stronger, even absolute, inhibiting effect. Thus, the researchers

Fig. 8.5 Inhibition of toxin
action in cell culture. The
effect of various amounts of
the single peptide (□), the
inhibitor (●) and the control
of the flexible backbone
alone (■). (**a**) inhibition of
the LF enzyme association
with the PA protein. (**b**)
Inhibition of diphtheria
toxin toxicity in cells

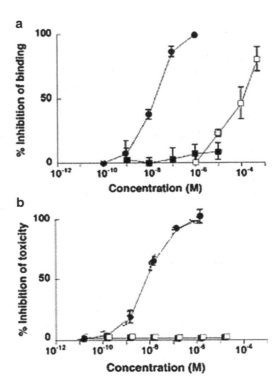

developed an inhibitor molecule that is constituted of 22 copies of the peptide
bound together to a synthetic material that served as a flexible molecular backbone.
To assess the effect of the new inhibitor comparing to the single peptide, they again
added each of them, separately, to hamster ovary cells, together with the constituents of the toxin. Assembly was assessed through radioactive protein tagging
(Fig. 8.5a).

The experiment demonstrated that the inhibitor interfered with the binding of the
PA protein with the LF enzyme much more efficiently and in much smaller
concentrations. In fact, the activity of the inhibitor was 7,500-folds greater than
that of the single peptide. A control substance (the flexible backbone without the
peptides) had absolutely no interfering effect on the assembly of the toxin.

Next, the researchers attempted to assess the activity of the synthesized inhibitor
in preventing intoxication. Ovarian cells were grown in culture in the presence of
the constituent proteins of the anthrax toxin to which a portion of the diphtheria
toxin was fused. The diphtheria toxin was aimed to serve as a surrogate marker:
once inside the cell it paralyzes cytoplasmic protein synthesis, an event that can be
easily observed and would mark successful assembly of the anthrax toxin subunits
and endocytosis into the cells. As shown in the graph (Fig. 8.5b), the inhibitor
prevented intoxication, while the control substance (the flexible backbone alone)
had no effect at all.

Table 8.2 Inhibition of anthrax toxin action in animal model

Inhibitor	Amounts (in nanomoles)			Outcome
	PA	LF	Peptide	
None	0.5	0.1	–	Symptoms of intoxication
Peptide + backbone	0.5	0.1	75	Symptoms of intoxication
Inhibitor	0.5	0.1	12	Delayed intoxication
Inhibitor	0.5	0.1	75	No symptoms

Table 8.2. All rats were anesthetized and injected with similar doses of the LF + PA solution, along with an additional substance depending on their assigned group (each group consisted of four rats): (1) Without an inhibitor the rats showed symptoms of intoxication 1 h after injection. (2) Free peptide + free backbone (not assembled together) showed symptoms 1 h after injection. (3) At small doses of the inhibitor, symptoms were delayed and presented 2 h after injection. (4) By injecting a high dose of the inhibitor, no symptoms appeared during a 1 week follow-up

The last test for the efficacy of the drug was performed by injecting the anthrax toxin and the inhibitor to living rats. The researchers used rat specie that is known to be highly sensitive to anthrax and dies from intoxication within a few hours only. The rats were injected with a comparatively large dose of the PA and LF proteins that are responsible together to the lethal symptoms of anthrax disease, and the therapeutic potential of the inhibitor was assessed (Table 8.2).

In the rat model, injection of the inhibitor together with the toxin delayed the symptoms of anthrax from appearing. A higher dose of the inhibitor prevented the symptoms of intoxication completely. The rats were protected even when the inhibitor was injected 3–4 min after the injection of the toxin. The control substance and the single inhibiting peptides had no effect on the evolution of symptoms in the rats. After the heroic saving from death, the safety of the drug itself was confirmed. In a 1 week follow-up of the rats, no side effects of the inhibitors were shown. The efficacy of the inhibitor in preventing the activity of the anthrax toxin in the animal model give hope that it, or another inhibitor developed by a similar approach, will supply an efficient therapy for anthrax.

Epistemology
The scientific information is presented as certain throughout the JRV article. No doubt in the finding can be identified in the text. The conclusions from the study state the developed drug "will supply an efficient therapy for anthrax".

References

Baram-Tsabari, A., & Yarden, A. (2005). Text genre as a factor in the formation of scientific literacy. *Journal of Research in Science Teaching, 42*(4), 403–428.
Mourez, M., Kane, R., Mogridge, J., Metallo, S., Deschatelets, P., Sellman, B., Whitesides, G., & Collier, R. (2001). Designing a polyvalent inhibitor of anthrax toxin. *Nature Biotechnology, 19*, 958–961.

Chapter 9
West Nile Virus: An Annotated Example of an Adapted Primary Literature (APL) Article

In this chapter we provide an example of an APL article annotated for epistemology and structure. The APL article originated from a Primary Scientific Literature (PSL) article (Wonham et al. 2004) and reports on the development of a mathematical model of the West Nile virus cross infection between birds and mosquitoes. The virus was first identified in the West Nile area of Uganda in the early first half of the twentieth century. It spread to parts of the Mediterranean and to Europe and turned up in North America in the late 1990s. People who get West Nile virus usually have been bitten by an infected mosquito, which typically has bitten an infected bird. The virus can lead to inflammation of the brain, which can be fatal. The key to fighting any disease is understanding its epidemiology, which often points to the most effective point of intervention. In order to prevent an epidemic of any infectious disease, attempts are made to keep the infection below some threshold in the infected populations. The APL article below reports the research of a group of mathematical biologists attempting to identify what that threshold might be for the West Nile virus, and when infected how to keep the infection below the epidemic threshold is most important.

Two other versions of this PSL article were developed: (i) a Journalistic Reported Version (JRV, Wonham 2004); and (ii) a web-based resource with supplemental pedagogical units, which were described in detail by Norris et al. (2009), and can be found at the web site of the University of Alberta, Edmonton, Canada (University of Alberta 2014) at http://www.edpolicystudies. ualberta.ca/CentresInstitutesAndNetworks/CRYSTALAlberta/Projects/Mathematical Modeling/WestNileVirusMathematicalModelingtoUnderstandandControlaDisease. aspx. The web-based resource was used and evaluated in two classrooms (Norris et al. 2009), and the APL version which appears below along with the JRV version were used and evaluated in high schools in Canada (Norris et al. 2012, see Chap. 7). Note that in the web-based version, key terms are highlighted and a click opens a pedagogical column which provides a more in-depth explanation, reminiscent in a way of the bolded terms and glossary included in the APL example which appears in

© Springer Science+Business Media Dordrecht 2015
A. Yarden et al., *Adapted Primary Literature*, Innovations in Science
Education and Technology 22, DOI 10.1007/978-94-017-9759-7_9

Chap. 8. The web-based version also includes questions for guiding deeper under-standing, similarly to the questions that accompany the two APL articles that are appear in the Appendices.

The PSL article, which served as the basis for the various adaptations described above, is composed of the following sections: Abstract, Introduction, Model description, Model analysis, Public health implications, Model predictions, and Temporal extensions (Wonham et al. 2004). These sections were partially retained in the APL version of the article which appears below, and the adapted article is composed of the following sections: Abstract, Introduction, Model development, Model simulations, Disease control, and Conclusions.

West Nile Virus: When Will an Outbreak Occur and How Can We Prevent It?[1]

Marjorie Wonham[*]
University of Alberta, Edmonton, AB, Canada

Gerda DeVries
Department of Mathematical & Statistical Sciences, University of Alberta, Edmonton, AB, Canada

Abstract

Emerging infectious diseases are a concern for wildlife and human health. To understand, prevent, and control infectious diseases, it can be helpful to use mathematical models. West Nile virus is a new infectious disease in North America. The virus is transmitted back and forth continually between mosquitoes and birds, and occasionally is transmitted from mosquitoes to humans. We developed a mathematical model to represent the West Nile virus outbreak in North America. We then used the model to determine whether it would be more effective to remove mosquitoes or remove birds to prevent disease outbreaks. Our model showed that reducing mosquito abundances would help prevent an outbreak. Reducing bird abundances, however, would have the opposite effect, increasing the chance of an outbreak.

---

[1] This paper is an adaptation based upon Wonham et al. (2004).

[*] This adaptation was completed when Marjorie Wonham was a postdoctoral fellow at the University of Alberta.

This paper offers a theoretical model of how bird and mosquito populations interact with each other. The model first is depicted as a flow chart with cells representing both infected and susceptible birds and mosquitos and arrows representing movement between cells. The flowchart is then cast into mathematical form, and the equations constitute a testable theory. The authors do not present data to test the theory, but rather how the theory can make testable predictions.

Introduction

Emerging infectious diseases are a growing aspect of global change. These diseases are major challenges for public health, agriculture, and wildlife management. Most infectious diseases of humans originate from other host animals, and many are transmitted by an insect vector such as a mosquito. Examples include malaria, yellow fever, and West Nile virus. Controlling host-vector diseases can be very difficult. This study focuses on the outbreak of the emerging infectious disease West Nile virus in North America.

A West Nile virus infection can be fatal for birds and mammals, including humans. Beginning in 1999, the virus spread across the continent, causing an unexpectedly high number of bird, horse, and human fatalities (Fig. 9.1). This surprising effect prompted a strong public health interest in understanding, preventing and controlling West Nile virus outbreaks.

Some scientific work has no clear application at the time it is conducted and no motivation other than the intrinsic interest of the scientist in knowing the answer to a question. The scientists and mathematicians conducting this study perhaps had the same sort of intrinsic motivations. However, the motivation they offer is built on a social concern with an unfamiliar disease (in North America) and no clear means of control. The implicit assumption in the justifications for conducting the study on a matter of public health is that the potential benefits of reducing sickness and death among birds, humans, and other mammals is a basic value that legitimates the research.

Since West Nile virus lives in mosquitoes and birds, it could conceivably be controlled by eradicating mosquitoes or eradicating birds. Which of these approaches would be more effective? We used a mathematical model to investigate how the reduction of bird and/or mosquito populations would affect a West Nile virus outbreak.

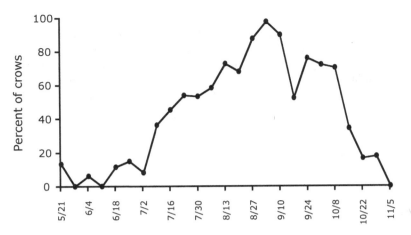

Fig. 9.1 Percent of dead American crows in New York City in 2000 that tested positive for West Nile virus (Data are redrawn from Bernard et al. 2001)

Organizational Structure

This paper is largely theoretical, and its organizational structure reflects that nature. This introductory section is primarily a justification for conducting the research. It describes a situation, raises questions that arise from that situation, and submits that it is important to seek answers to the questions. Much research, both theoretical and empirical, contains an introductory section organized similarly. The following section on model development is the longest and most important. It is subdivided in order to describe different elements of the model separately. The final four sections, beginning with model simulations, are about possible applications of the model to the prevention of epidemics. These final sections provide direction for rigorous testing of the model but do not report on any such tests that have been completed.

Goal-Directed Structure

The organizational structure serves the goal-directed structure of the paper, which is to provide a useful model for predicting the course of a West Nile virus outbreak under various conditions of mosquitos and bird control. That goal is stated in the previous paragraph.

Model Development

We based our model on the classic infectious disease model initially developed by Kermack and McKendrick (1927), and subsequently modified for host-vector diseases by Anderson et al. (1992). In this approach, the model keeps track of the number and fate of hosts and vectors that are and are not infected with the virus.

Objectivity
Note in the previous paragraph and in many subsequent ones the use of the first person, active voice, in which the subject of the sentence is also the agent of the action. Sometimes scientific writing uses the passive construction in which the subject is the recipient of the action. Moreover, the agent is often disguised, as in the following: "The model was based on the classic infectious. . ." instead of what appears in the first sentence of the paragraph. Some people believe that the first person active appears too subjective and thus prefer the passive voice. Of course, this switch in mode of expression has no bearing at all on the quality of the science, including its objectivity.

Birds

First, we considered the bird population in New York City during the 2000 West Nile outbreak. All the birds in that population can be classified into one of two groups: those that were infected with West Nile virus, and those that were not. (The third group of birds, the dead ones, doesn't influence the disease dynamics.) In epidemiological terminology, the birds that have West Nile virus are called "infectious", and those that don't have it are called "susceptible". So, our model needed to keep track of the number and fate of infectious and susceptible birds.

We visualized these two bird groups in a flow chart, as shown in Fig. 9.2. One box represents the number of infectious birds, abbreviated I_B, and another box the number of susceptible birds, abbreviated S_B. In this standard terminology, I and S refer to infectious and susceptible, respectively, and the subscript B indicates that these are birds. If a bird gets infected, it is subtracted from the susceptible group and added to the infectious group, as indicated by the arrow. Since all infectious birds die shortly after becoming infected, we added another arrow allowing birds to be subtracted from the infectious group.

Because we restricted our model to a single summer, we assumed that any births that year had already taken place earlier in the spring, and that the natural bird death rate was negligible over that time period, so we omitted bird births and natural deaths from the model.

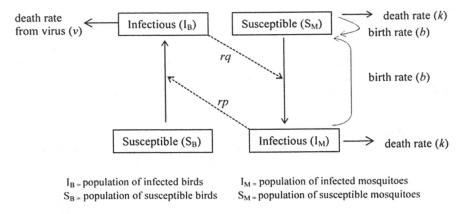

Fig. 9.2 Model for West Nile virus

Fallibility

Note the two explicitly stated assumptions in the previous paragraph. The assumptions themselves are not justified by any reason provided in the paragraph. The assumptions are used as the justification for something else, namely the omission of bird births and natural deaths from the model. If the assumptions are incorrect, these omissions would reduce the accuracy of the model. Presumably, the authors thought that the assumptions were close enough to the truth to not adversely affect the model.

Mosquitoes

Next, we considered the mosquito population. Again, all the mosquitoes in New York City can be divided into the infectious ones that carried West Nile virus, I_M, and the susceptible ones that did not, S_M. If a mosquito contracts the virus, it is subtracted from the susceptible group and added to the infectious group, as shown by the arrow.

Unlike for birds, we had to consider the mosquito birth and death rates. A female mosquito's lifespan is approximately 33 days, during which time she may lay approximately 10 egg masses, each of which takes approximately 15 days to mature into biting adults. During a summer season, a mosquito population would therefore move through multiple generations. To account for this process, we allowed both susceptible and infectious mosquitoes to give birth to new susceptible individuals. The birth rate is abbreviated b. We also allowed mosquitoes to be subtracted from both groups by dying, and the death rate is abbreviated as k.

Bird-Mosquito Interaction

The final step was to consider how the bird and mosquito populations interact with each other. A bird can be infected if it is bitten by an infectious mosquito. Likewise, a mosquito can be infected if it bites an infectious bird. In this way, the infection is transmitted from birds to mosquitoes to birds, and so on.

To represent these interactions, we added two dashed arrows. The lower arrow indicates that infectious mosquitoes can infect birds, and the upper one indicates that infectious birds can infect mosquitoes. Both arrows are labelled with the mosquito biting rate, r, and the probabilities of disease transmission, p and q.

Restrictions and Assumptions

To help keep our West Nile virus model simple and practical, we restricted it to treat certain species and a particular time frame, as follows:

1. We restricted the model to bird hosts and mosquito vectors. Although the West Nile virus spreads occasionally to other vertebrates, including humans, it seems generally not to return from them to mosquitoes.
2. We developed the model around specific mosquito and bird species: the dominant North American vector mosquitoes *Culex pipiens*, and the bird species with the best available infection and mortality data, the American crow *Corvus brachyrhyncos*.
3. We confined the model to represent a single north-temperate summer season, and validated it by comparing it to the best available dataset at the time, the New York City outbreak of 2000.

Fallibility

The previous three restrictions make the model less general. The reduction in generality must be offset by other benefits. The only justification given for the restrictions is that they would "help keep our West Nile virus model simple and practical." Why might simplicity and practicality be valued in science? Simplicity often is valued if it means that a considerable amount of a complex phenomenon can be understood by taking into account a relatively small number of factors. Practicality is valued if it means that a model is usable, or that the effort in using a model is worth the benefits that it provides. If a model is so complicated that it requires horrendously difficult mathematics, for example then, no matter how great its accuracy, the model might not be used. Concessions to practicality typically are accompanied by redirections in truth and generality, which is one factor leading to the fallibility of science.

We also made some simplifying assumptions to keep the model as straightforward as possible.

Fallibility

The following series of simplifying assumptions is extremely interesting. The justification provided is that they would make the model "as straightforward as possible". Yet, the effect the assumptions had on the model was varied. By admission, the scientists knew that assumptions 2 and 3 were false. The effect of acting on them would be to make the model misrepresent reality at least to some degree. Presumably, the justification for making the assumptions was that the misrepresentation was compensated by the increased straightforwardness of the model. Thus, on occasion, practical matters of working with and using the model could outweigh substantive matters such as theoretical verisimilitude. The scientists knew that the omission from the model noted in 6 would make the model less accurate, but by how much was not mentioned. The remaining five assumptions had unknown or unreported truth values. Thus, the effects on the veracity of the model of making these assumptions are unknown. Again, presumably, the scientists either believed that the effects were of so little consequence as to warrant little attention, or thought that negative outcomes produced by waiting for new research to clarify the status of the assumptions would be outweighed by the positive outcome of producing a model without delay.

1. Since birds live much longer and reproduce more slowly than mosquitoes, we assumed their per capita birth and death rates were negligible for a single season and therefore left them out of the model.
2. We assumed that all infectious mosquitoes give birth to uninfected offspring. Laboratory work has shown that virus transmission occurs at the extremely low level of approximately 0.1 %.
3. We assumed that the mosquito population remained constant all summer even though it may fluctuate during the summer according to temperature and precipitation.
4. We assumed that a mosquito carrying West Nile virus cannot lose its infection. We don't know for certain if this is true.
5. We assumed that an infected mosquito does not die of the virus, and is not in any other way affected by it.
6. We omitted two particular aspects of the mosquito life cycle and infection cycle. (1) Mosquitoes give birth to aquatic larvae, which require approximately 15 days to mature into biting adults. (2) Laboratory work shows that a mosquito that contracts West Nile must incubate the virus for approximately 8–12 days before it can be transmitted to a bird.

7. We assumed that birds could get infected with West Nile virus only by being bitten by an infectious mosquito.
8. We assumed that a hungry mosquito could always find a bird to bite.

Model Variables and Parameters

In mathematical modeling terminology, S_B, I_B, S_M, and I_M are known as variables. Since we were interested in how the values of these variables change over time, we defined time as an additional variable, denoted t. The fate of individuals in each group is described by parameters, which are indicated in Fig. 9.2 by arrows. We estimated the values of the parameters from published biological literature on birds and mosquitoes.

For birds, we needed to estimate the parameter v, the death rate from the virus, which we did as follows. An infectious crow dies from West Nile virus in approximately 5 days. In other words, an infectious crow has an approximate probability of $1/5 = 0.20$ of dying every day. We can therefore say 0.20 is the approximate proportion of the infectious crow population that dies every day. This is known as the average per capita daily death rate for infectious crows, abbreviated v.

We used similar calculations for the mosquito birth and death rate parameters. The average lifespan for an adult female *Culex pipiens* mosquito is approximately 33 days. Thus, an individual mosquito has a daily probability of dying of $1/33 = 0.03$. This is the mosquito average per capita daily death rate, which we abbreviated k.

Since we assumed, for simplicity, that the mosquito population remains constant over the summer, we set the daily per capita mosquito birth rate, abbreviated b, equal to the death rate, k.

The parameters associated with the dashed arrows in Fig. 9.2 required a little extra consideration. In general, it seemed reasonable that West Nile virus transmission should depend on the mosquito biting rate, the proportion of bites that actually transmit the disease, and the relative numbers of mosquitoes and birds. An adult female mosquito lives in a continuous cycle of feeding, laying eggs, feeding, laying eggs, and so on. It takes approximately 2.3 days for a mosquito to convert a blood meal into a batch of eggs, so the maximum biting rate for an individual mosquito is approximately once every 2.3 days. This gives an average daily per capita biting rate of $1/2.3 \approx 0.44$ bites per mosquito per day, denoted r. By assuming that mosquitoes were biting at their maximum rate, our model used so-called frequency-dependent transmission dynamics.

The remaining two parameters were disease transmission probabilities. From laboratory work, we knew that not every bite by an infectious mosquito on an infectious bird leads to virus transmission. From infectious mosquitoes to birds, the transmission probability is ≈ 0.88 (denoted p). From infectious birds to mosquitoes, it is only ≈ 0.16 (denoted q).

Argumentative Structure

The previous two paragraphs contain arguments the conclusions of which are values for parameters in the model. In each case, the argument begins with facts derived from previous research (e.g., "adult female mosquito lives in a continuous cycle of feeding, laying eggs, feeding, laying eggs"; "the maximum biting rate for an individual mosquito is approximately once every 2–3 days"). From these facts logical arguments lead to parameter values.

Mathematical Equations

We converted the flow chart diagram in Fig. 9.2 into a system of mathematical equations. There is one ordinary differential equation for each group, or variable.

The equation for the rate of change in the number of susceptible birds is:

$$\frac{dS_B}{dt} = -rpI_M\frac{S_B}{N_B} \tag{9.1}$$

Equation 9.1 tracks the change in the number of susceptible birds over time. The expression on the left-hand side, dS_B/dt, is the derivative of S_B with respect to t, or the instantaneous rate of change in the number of susceptible birds. The right-hand side is made up of model parameters and variables.

The parameter r is the per capita daily mosquito biting rate: the number of bites per mosquito per day. The parameter p is the probability that a bite by an infectious mosquito will infect a bird. Thus, the product rp is the per capita disease-transmitting biting rate: the number of disease-transmitting bites per infectious mosquito per day. The variable I_M is the number of infectious mosquitoes. Thus, the product rpI_M is the total disease-transmitting biting rate: the number of disease-transmitting bites per day across the entire mosquito population.

The fraction S_B/N_B is the proportion of birds that are susceptible. Thus, the product rpI_MS_B/N_B is the complete disease-transmitting biting rate: the number of disease-transmitting bites per day by infectious mosquitoes on susceptible birds. In other words, it is the daily rate at which susceptible birds become infected. The term rpI_MS_B/N_B is negative in Eq. 9.1, because infected birds are removed from the susceptible group and added to the infectious group.

Argumentative Structure

The preceding four short paragraphs serve two purposes: to describe Eq. 9.1 and its meaning and to argue that the equation captures the rate of change in the number of susceptible birds.

The equation for the rate of change in the number of infectious birds is:

$$\frac{dI_B}{dt} = rpI_M\frac{S_B}{N_B} - vI_B \tag{9.2}$$

Equation 9.2 tracks the change in the number of infectious birds over time. The expression on the left-hand side, dI_B/dt, is the derivative of I_B with respect to t, or the instantaneous rate of change in the number of infectious birds.

The right-hand side is made up of model parameters and variables, and describes the instantaneous rate of change of the number of infectious birds.

The first term on the right-hand side, rpI_MS_B/N_B is the same as the term on the right-hand side of Eq. 9.1. The term is negative in Eq. 9.1, since infected birds are removed from the susceptible group, and positive in Eq. 9.2, since infected birds move into the infectious group. The second term is vI_B. The parameter v is the per capita daily death rate caused by West Nile virus infection, or the proportion of the infectious bird population that dies every day. The product vI_B is therefore the total number of infectious birds that die from the virus every day. This term is subtracted, because the number of infectious birds is decreased when some of them die. When these two terms are taken together, we can see that adding the rate at which new infectious birds arise and subtracting the death rate gives the rate of change in the number of infectious birds.

Argumentative Structure
The structure and intent of the previous three paragraphs is similar to those related to Eq. 9.1. The aim is to describe Eq. 9.2 and to justify it as an accurate representation of the rate of change in the number of infectious birds.

$$\frac{dS_M}{dt} = bS_M - kS_M - rqS_M\frac{I_B}{N_B} \tag{9.3}$$

Equation 9.3 tracks the change in the number of susceptible mosquitoes over time. The expression on the left-hand side is the instantaneous rate of change in the number of susceptible mosquitoes.

On the right-hand side, the first term is bS_M. The parameter b is the per capita daily mosquito birth rate. The product bS_M is therefore the total number of new mosquitoes produced per day. This rate is positive, since new mosquitoes are added to the susceptible group. The second term is kS_M. The parameter k is the per capita daily mosquito death rate. The product kS_M is therefore the total number of mosquitoes that die each day, or the total death rate. This term is subtracted, since dead mosquitoes move out of this group. The first two terms together, bS_M kS_M, give the net rate of change in the number of susceptible mosquitoes due to births and deaths.

The third term is rqS_MI_B/N_B. The parameter r is the per capita daily mosquito biting rate: the number of bites by a mosquito each day. The parameter q is the probability that biting an infectious bird will infect a mosquito. Thus, the product rq is the per capita disease-contracting biting rate: the number of disease-contracting bites per mosquito per day. The variable S_M is the number of susceptible mosquitoes. Thus, the product rqS_M is the total disease-contracting biting rate: the number of bites per day that lead to the disease being contracted by susceptible mosquitoes. The fraction I_B/N_B is the proportion of birds that are infectious. Thus, the product rpS_MI_B/N_B is the complete disease-contracting biting rate: the number of disease-contracting bites per day by susceptible mosquitoes on infectious birds. In other words, it is the daily rate at which susceptible mosquitoes become infected.

The term rpS_MI_B/N_B is subtracted in Eq. 9.3, because infected mosquitoes are removed from the susceptible group. When these three terms are taken together, we can see that adding the birth rate, subtracting the death rate, and subtracting the rate at which new infectious mosquitoes arise gives the total rate of change in the number of susceptible mosquitoes.

$$\frac{dI_M}{dt} rqS_M\frac{I_B}{N_B} - kI_M \qquad (9.4)$$

Equation 9.4 tracks the change in the number of infectious mosquitoes over time. The expression on the left-hand side, is the instantaneous rate of change in the number of infectious mosquitoes.

On the right-hand side, the first term rqS_MI_B/N_B is the same as the third term on the right-hand side of Eq. 9.3, though it is positive in Eq. 9.4 because infected mosquitoes move into the infectious group.

The second term is kI_M. The parameter k is the per capita daily mosquito death rate. The product kI_M is therefore the total number of mosquitoes that die each day, or the total death rate. This term is subtracted, since dead mosquitoes are removed from this group.

When these two terms are taken together, we can see that adding the rate at which new infectious mosquitoes arise and subtracting the death rate gives the total rate of change in the number of infectious mosquitoes.

The discussion of the preceding several paragraphs referring to Eqs. 9.3 and 9.4 provide descriptions of the meanings of the equations and arguments that the equations represent the reality they are meant to capture. At this point the model has been completely described and justified and is ready to be implemented to see what it predicts about the control of a West Nile virus outbreak.

Argumentative Structure

The argumentative structure of the article is displayed in the section on model development and in the final four sections on application. The arguments differ in form. On the model development section, the overall argument is logical-mathematical. Basically the argument structure is: "If we make such-and-such assumptions about bird and mosquito interactions, then such-and-such mathematical equations describe those interactions." The arguments exist in creative constructions.

In the four final sections, the arguments make predictions from the developed model about what would occur or would have occurred under various real-world scenarios. The arguments are deductions from the model and the assumed real-world conditions.

Model Simulations

In order to visualize the model's predictions, we needed to simulate the model. To do this, we gave each parameter its average value and we gave each of the variables, S_B, I_B, S_M, and I_M, an arbitrary initial value.

We then used a computer program to calculate the changes in each variable over time, for 120 days. The simulation results are visualized on graphs showing how the numbers of susceptible and infectious birds and mosquitoes are predicted to change over time (Fig. 9.3a, b).

We validated the model by comparing the simulations to observed data, and found that its predictions were sufficiently realistic to be useful. Based on this result, we continued to investigate the model to determine how best to control a potential West Nile outbreak.

Fallibility

Interesting language was chosen in the previous paragraph to evaluate the outcomes of the model validation. The criteria used are realism and usefulness. Saying that the model is realistic is perhaps saying something about how accurately it reproduces the observed data. Saying that it is "sufficiently realistic to be useful" suggests that the model contained inaccuracies but that these were small enough that the model could still be used. The model is at least better than nothing and perhaps a good deal better.

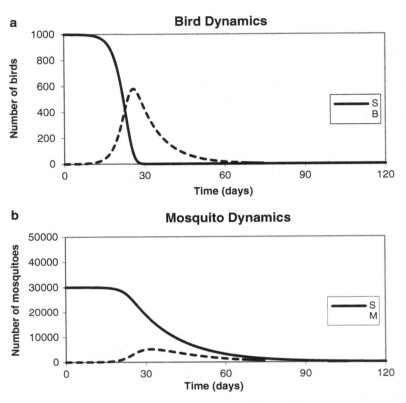

Fig. 9.3 (a) This panel shows changes in the abundance of the two bird groups, susceptible (S_B) and infectious (I_B). (b) This panel shows changes over time in the abundances of the two mosquito groups over 120 days following the introduction of West Nile virus to the bird-mosquito system

Disease Control

Disease Reproduction Number, R_0

R_0 ("R-zero" or "R-naught") is defined as the number of new infections that would result from the introduction of a single infectious individual (bird or mosquito) into an entirely susceptible population of birds and mosquitoes.

What does the reproduction number mean in practice? If the value of R_0 is less than one, it means that one infectious individual would generate, on average, fewer than one new infection. In this case, the disease would die out. On the other hand, the value of R_0 could be greater than one. In this case, an infectious individual would generate, on average, more than one new infection, and a disease outbreak would occur.

R_0 is calculated from the equations for the disease model, and is an expression composed of the model's parameters and the initial values of the variables. In many cases, these parameters and initial values can be manipulated by humans to increase or decrease R_0. Thus, examining the expression for R_0 can help us figure out the best way to try to prevent a disease outbreak.

For our model, the R_0 calculation gives the expression:

$$R_0 = \sqrt{\frac{rp}{k} \frac{rqN_{M0}}{vN_{B0}}} \tag{9.5}$$

Here, the subscript 0 in N_{M0} and N_{B0} indicates the initial values of these variables, on day 0. (In contrast, the subscript 0 in R_0 does not indicate day 0—it is just the conventional abbreviation for the disease reproduction number.) The expression on the right-hand side is made up of model parameters and the initial values of two of the variables. It consists of two fractions under a square root sign.

The first fraction is rp/k. The parameter r is the per capita daily mosquito biting rate: the number of bites by a mosquito each day. The parameter p is the probability that a bite by an infectious mosquito will infect a bird. Thus, the product rp is the per capita disease-transmitting biting rate: the number of disease-transmitting bites per infectious mosquito per day. The parameter k is the per capita daily mosquito death rate: the chance a mosquito has of dying each day. (In other words, $1/k$ is the mosquito's average lifespan.) The first fraction can thus be read as the number of birds an infectious mosquito can bite and infect each day (rp), multiplied by the number of days the mosquito lives ($1/k$). In other words, it is the number of bird infections caused by a single infectious mosquito before it dies.

The second fraction is composed of the two factors rq/v and N_{M0}/N_{B0}. In the first factor, the parameter r is the per capita daily mosquito biting rate: the number of bites by a mosquito each day. The parameter q is the probability that biting an infectious bird will infect a mosquito. Thus, the product rq is the per capita disease-contracting biting rate: the number of disease-contracting bites per mosquito per day. The parameter v is the per capita daily bird death rate caused by West Nile virus. In other words, it is the chance a bird has of dying each day. Therefore, $1/v$ is the average lifespan of an infectious bird. This factor can thus be read as the number of mosquitoes an infectious bird can infect each day (rq), multiplied by the number of days the bird lives ($1/v$). In other words, it is the number of mosquito infections caused by a single infectious bird before it dies.

In the second factor, the variable N_{M0} is the initial number of mosquitoes and the variable N_{B0} is the initial number of birds. This factor appears in this term because the number of mosquitoes that a bird can infect depends on the number of mosquitoes biting that bird, which depends of the number of mosquitoes per bird, N_{M0}/N_{B0}.

This past section has provided a detailed description of how R_0 is constructed and what each term in the expression means. Some logical reasoning to conclusions is found, all in the service of articulating the meaning of R_0.

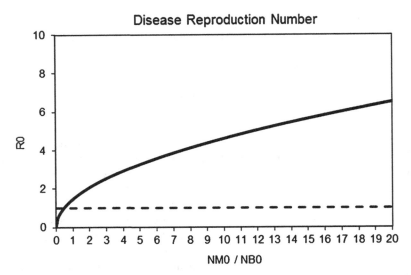

Fig. 9.4 Disease reproduction number as a function of the ratio of the initial number of mosquitoes to the initial number of birds

Using R_0

Even without calculating the value of R_0, the structure of Eq. 9.5 gives us some insight into how to prevent a West Nile virus outbreak. We know that higher R_0 values (above one) mean a greater chance of an outbreak and lower R_0 values (below one) mean a lesser chance of an outbreak. Simply by looking at the expression for R_0, we see that the mosquito abundance is in the numerator and the bird abundance in the denominator. This tells us that decreasing the mosquito abundance would decrease R_0, and decrease the chance of an outbreak. It also tells us that decreasing the bird abundance would have the opposite effect: it would increase R_0, and increase the chance of an outbreak. We visualize this relationship by plotting R_0 as a function of the relative numbers of mosquitoes and birds (see Fig. 9.4).

The solid line plots R_0 and the dashed line plots the threshold value of one, above which a disease outbreak will occur and below which it will not. The expression for R_0 allows us to calculate the threshold number of mosquitoes per bird that would lead to a West Nile virus outbreak. We determine this number by setting R_0 to its threshold value of one, and rearranging Eq. 9.5 to solve for the threshold value of the ratio N_{M0}/N_{B0}. Above this threshold, a disease outbreak would be predicted to occur. Below the threshold, it would not. We found that a 40–70 % reduction in the mosquito population would have prevented the outbreak in New York City in 2000.

Argumentative Structure

This past section describes how R_0 can be used to prevent a disease out-break. The argument first made is that reducing the mosquito population reduces the chance of a disease outbreak and that reducing the bird population increases the risk. The second argument is that reducing the mosquito population by 40–70 % would have prevented the 2000 outbreak in New York City. Note that the inexactness captured by the range of 30 % speaks to the uncertainty inherent in the model.

Conclusions

Is it more effective to control West Nile virus by reducing birds or by reducing mosquitoes? By constructing and analyzing a model of West Nile virus, we were able to answer this question. We hope that results from this model will be useful in designing strategies to control West Nile virus in specific locations. However, we expect that the model's parameter values will differ from place to place. Therefore, before the model can be applied to a given locale, biological research must be done to determine accurate parameter values for that place. In future work, we would like to build a series of different models for West Nile virus to determine which of our assumptions and restrictions make the model generate less or more accurate predictions.

References

Anderson, R. M., May, R. M., & Anderson, B. (1992). *Infectious diseases of humans: Dynamics and control*. New York: Oxford University Press.

Bernard, K. A., Maffei, J. G., Jones, S. A., Kauffman, E. B., Ebel, G., Dupuis, A. P., 2nd, Ngo, K. A., Nicholas, D. C., Young, D. M., Shi, P. Y., Kulasekera, V. L., Edison, M., White, D. J., Stone, W. B., Kramer, L. D., & NY State West Nile Virus Surveillance Team. (2001). West Nile virus infection in birds and mosquitoes, New York State. *Emerging Infectious Diseases, 7* (4), 679–685.

Kermack, W. O., & McKendrick, A. G. (1927). A contribution to the mathematical theory of epidemics. *Proceedings of the Royal Society of London A, 115*, 700–721. doi:10.1098/rspa.1927.0118.

Norris, S. P., Macnab, J. S., Wonham, M., & de Vries, G. (2009). West Nile virus: Using adapted primary literature in mathematical biology to teach scientific and mathematical reasoning in high school. *Research in Science Education, 39*(3), 321–329.

Norris, S. P., Stelnicki, N., & de Vries, G. (2012). Teaching mathematical biology in high school using adapted primary literature. *Research in Science Education, 42*(4), 633–649.

University of Alberta. (2014). *West Nile virus: Mathematical modeling to understand and control a disease CRYSTAL-Alberta*. Available: http://www.edpolicystudies.ualberta.ca/Centres Institutes AndNetworks/CRYSTALAlberta/Projects/MathematicalModeling/WestNileVirusMathematical ModelingtoUnderstandandControlaDisease.aspx. October 27 2014

Wonham, M. (2004). The mathematics of mosquitos and West Nile virus. *Pi in the Sky, 8*, 5–9.

Wonham, M. J., deCamino-Beck, T., & Lewis, M. A. (2004). An epidemiological model for West Nile virus: Invasion analysis and control applications. *Proceedings of the Royal Society of London. Series B: Biological Sciences, 271*(1538), 501–507. doi:10.1098/rspb.2003.2608.

Chapter 10
Coronal Heating: An Annotated Example of an Adapted Primary Literature (APL) Article

In this chapter we provide an example of an APL article annotated for epistemology and meta-scientific language. The APL article originated from a Primary Scientific Literature (PSL) article (Aschwanden et al. 2007), and reports on the Sun's coronal heating problem. This problem refers to the puzzling fact that the Sun's corona is much hotter than the lower layers of the Sun, which lie closer to its energy-producing core. Over the last six decades, hundreds of theoretical models to explain the corona's high temperature have been proposed. There is still no obvious solution in sight, partly because many difficulties arise in trying to understand why the corona is so hot. The original PSL article discusses ten pieces of observational evidence to support the two-step heating scenario.

The APL article below considers the evidence for a series of proposed solutions to the coronal heating problem. The article offers arguments for why several of these solutions must be rejected and indeed argues that the term 'coronal heating' is a misnomer, because the heating does not take place in the corona. The paper demonstrates a genuinely controversial issue in astrophysics (at the time the original article was published) and illustrates how scientists go about trying to resolve controversy.

The PSL article, which served as the basis for the adaptation, is composed of the following sections: Abstract, Introduction, Coronal heating requirement, Ten arguments for heating in the chromospheres/transition region, and Conclusion. These sections were retained in the APL version of the article which appears below, but instead of ten it provides three arguments for heating in the chromospheres/transition region.

© Springer Science+Business Media Dordrecht 2015
A. Yarden et al., *Adapted Primary Literature*, Innovations in Science Education and Technology 22, DOI 10.1007/978-94-017-9759-7_10

The Coronal Heating Problem[1]

Adrienne Parent[*]
Department of Physics, University of Alberta, Edmonton, AB, Canada

Abstract

The "coronal heating problem" refers to the puzzling fact that the Sun's corona is much hotter than the lower layers of the Sun, which lie closer to its energy-producing core. It would be shocking to find that as you moved away from a campfire the air became many times hotter than the fire itself, but this is similar to what happens in the case of the Sun. Over the last six or more decades, hundreds of theoretical models to explain the corona's high temperature have been proposed. There is still no obvious solution in sight, partly because many difficulties arise in trying to understand why the corona is so hot. For example, scientific models, which propose various heating processes that might solve the coronal heating problem, rely on being proven true or false by observations of the Sun. The best observations of the solar corona are performed by instruments onboard spacecraft that fly outside of the Earth's atmosphere, but spacecraft are incredibly expensive and take many years to develop and build. Only in recent decades has their use in studying the Sun become more practical. Despite these and other difficulties, we re-examine the coronal heating problem in this paper. We point out that spacecraft observations of the Sun show no evidence that the actual heating is occurring locally within the Sun's corona. Instead, observations suggest that the heating occurs in a two-step process. First, the gas-like material that makes up the Sun, called plasma, becomes heated below the corona in the lower layers of the Sun's atmosphere. Second, the plasma moves up into the corona along curved paths called coronal loops. Three pieces of new observational evidence for this two-step heating scenario are discussed in the paper: (1) the temperature evolution of coronal loops, (2) the overdensity of hot coronal loops, and (3) upflows in coronal loops. By thinking about the coronal heating problem in terms of the two steps described above, it is possible to narrow down the number of theoretical models that could explain coronal heating in active and quiet regions of the Sun. Note, however, that our

---

[1] This paper is an adaptation based upon Aschwanden et al. (2007) and Aschwanden (2001). Additional information was provided by the following text: Aschwanden (2004). We thank Dr. Markus J. Aschwanden for his permission to adapt his scientific papers and textbook material for this project. His enthusiasm and willingness to share good quality images from his publications are very much appreciated.

[*] Adrienne Parent, authored the following adaptation as part of a research assistantship during her doctoral studies in Physics.

arguments do not apply to coronal holes and those parts of the corona which extend far away from the Sun, out into the heliosphere.

This paper is an extended scientific argument. Even so, it provides only three of the ten arguments found in the original work. The paper presents no new data, but draws upon existing data to argue for a favoured hypothesis and to impeach alternative hypotheses. The *Astrophysical Journal* publishes only original research, so in this branch of science a new argument based upon existing data qualifies as original.

Introduction

In the early 1940s, physicists Bengt Edlén and Walter Grotrian discovered the true source of strange emission lines in the spectrum of light from the Sun's corona. The emission lines had previously been attributed to a mysterious element, called "coronium," that was thought to exist only on the Sun, but Edlén and Grotrian found that they were in fact due to the presence of iron and calcium atoms that had been stripped of many of their electrons (9–13 in the case of iron). For these atoms, the loss of so many electrons meant that the temperature of the Sun's corona had to be at least 1 million kelvins (1 MK). This was a surprising result, since the temperature of the photosphere below was known to be only about 6,000 K. How can it be that the temperature of a hot, radiating object like the Sun increases as you move away from its center, instead of decreasing as we would naturally expect?

Rationality
In science, paradoxical findings such as the one described in the preceding paragraph often motivate research. In an epistemology based on rationality, a paradox cannot be left unaddressed because contradictions, or seeming con-tradictions, challenge reason. Thus, the paradox gives rise to a scientific question. ... "How can it be...?"

The counterintuitive fact that the Sun's corona is more than 200 times hotter than its lower boundary, the chromosphere, has puzzled solar physicists for over 60 years. The attempt to solve the coronal heating problem by determining and understanding the dominant heating processes in the solar corona has proved to be a difficult task. Many theoretical heating models have been developed by physicists, but difficulties arise because the ultimate test for any heating theory is whether its predictions match observations of the Sun's corona made by instruments onboard spacecraft.

Rationality

Scientists regard it as insufficient for theoretical models to explain the data that motivated their development. Although necessary, this is too easy a test to pass. This is the point the authors are implying in the previous sentence when they say that the ultimate test for a theory is successful predictions. It is at this point that fallibility comes to the fore, and scientists realize how easy it is to be wrong. Thus, they look for new findings that the model predicts and check for their presence by subsequent observations.

One such difficulty involves the spacecraft technology itself. Only since the 1960s has the technology been available to send ships into space to observe the Sun from outside the Earth's atmosphere. Even with adequate technology in place, the design, implementation and launch of a spaceship make up an expensive and high-risk effort. Therefore, spacecraft missions focused on observing the Sun have not been numerous and the data required to test theories are not plentiful. In addition, some theories are based upon parameters that cannot be measured by existing spacecraft instruments and will not be measurable until scientists create new instrument technology, so these theories currently cannot be tested. Finally, the process of testing theory against experiment is made even more complicated by the fact that available observational data from spacecraft can often be interpreted in different ways. Thus, the acceptance or rejection of theories becomes the subject of scientific debate in the solar physics community and definitive answers to the coronal heating problem are delayed until new observations become available.

Fallible Rationality

In the previous paragraph the author has described the burdens that fallible rationality brings. Conclusions can come slowly, and they can demand an enormous commitment in time, effort, and money. Moreover, the data often point in more than one direction, so that multiple, incompatible conclusions are common. Being prepared to reach a conclusion only when the evidence is sufficient, recognizing all the while that even such a long-awaited conclusion can be incorrect, makes the scientific process deliberate and is precisely why it can appear needlessly tedious and painstaking to the non-scientist.

Consider a house that is heated by a centralized heating source, for example, a furnace in the basement. In this house, the heat generated in the basement is then distributed throughout the building via air ducts or water pipelines, and this is how the house stays warm. We would like to argue that in a similar way the hot temperature of the Sun's corona is generated below the corona by a primary heating process located in the solar transition region or upper chromosphere. That is, the heat generation in the case of the Sun does not occur in the corona itself, but rather in the chromosphere below. The hot plasma from the chromosphere is then distributed upward throughout the corona along coronal loops, causing the corona to become very hot.

Goal-Directedness

In the previous paragraph the author asks us to consider an analogy, namely that heating in the Sun's corona can occur through a process similar to the production of heat in a basement furnace of a house and the subsequent distribution of the heat throughout the house by way of ductwork or piping. In science, an analogy is often used, as it is in this case, to clarify a hypothesized model for producing a phenomenon. Although the analogy may add to the plausibility of the model it is not really part of the reasons supporting the model. Rather, the plausibility is used more rhetorically as a means to maintain interest in the model while its implications and means of testing are being worked out.

Returning to the idea of a house, it could be said, however, that a centralized heating source in the basement is not required for the house to maintain a warm temperature throughout its rooms. Instead, the heating of the house could be accomplished by an external source. For example, sunlight shining down on the building's windows could cause the air inside to warm up. In a similar way, the Sun's corona could be heated by external waves that actually originate from below the Sun's atmosphere entirely, closer to the core. We will see, however, that such a scenario cannot explain many observations and so lacks observational support. Therefore, if we accept that the primary heating occurs below the corona in the chromosphere and/or transition region, with the hot plasma being distributed upwards and throughout the corona along loops, then the phrase "coronal heating process" becomes a poor name for the whole procedure. The hot temperature of the corona is not caused by heating processes within the corona itself, but is caused by upflows of heated plasma from below.

Goal-Directedness

In the preceding passage the author anticipates an argument against the model and provides a sketch of how the argument will be countered and defeated. The author also states clearly and forthrightly the main conclusion of the paper. Signaling in ways such as these what is to come in the paper is a means of alerting the reader and focusing the reader's mind in the direction preferred by the author. Up to this point, although the conclusion has been stated, no evidence or reasons have been advanced in its favour. The author's hope is that the reader now is primed to read the support and is anticipating its appearance.

Coronal Heating Requirement

In order for the corona to maintain its million-degree temperature, heat energy must constantly be supplied to it at a rate that matches how fast energy is lost by the corona. If this balance between energy transmission rates were not maintained, then either the corona's overall temperature would continue to grow, or it would continuously cool down, but these cases are not observed. It is clear then that the corona requires a certain amount of energy to stay hot, and it is important to identify which regions of the Sun's atmosphere contribute the most heat energy to the overall energy requirement of the corona. Heating processes that occur in the regions that contribute the most energy to the corona become the most important processes for overall coronal heating. Once we identify these important heating processes, we are on our way to better understanding, and maybe even solving, the coronal heating problem.

> **Argumentative Structure**
> This last paragraph presents the author's version of the logic of the coronal heating problem. It is the rational case motivating interest in the phenomenon: The corona's temperature is neither falling nor rising (within some unstated temperature limits). A hot substance like the Sun's corona radiates energy to other objects and into space. An object that is radiating energy but remaining at a constant temperature must have a source of energy input. From what source does the corona receive its energy?

To identify the regions of the Sun that contribute the most heat to the corona, it is helpful to examine a soft X-ray image of the Sun, shown in Fig. 10.1 (top panel), that has been obtained with the Soft X-Ray Telescope (SXT instrument) onboard the Yohkoh spacecraft. In this picture, the X-ray emissions from the extended corona have been captured out to a distance of about two solar radii from the center of the Sun. Just as we humans use light in the visible wavelengths to observe regular objects around us, X-ray emissions recorded by special telescopes can show us what the plasma in the corona looks like, since the plasma radiates X-rays when it has temperatures exceeding about 1 MK. The power per unit area, F_H, required to heat each of the 36 sectors (i.e., the energy required per second and per m^2) can be estimated from a careful analysis of the brightness of the corona in the soft X-ray Yohkoh image. On the image, the solar corona above the edge of the Sun's "surface" (marked by the inner circle) has been divided into 36 sectors with 10° width each. This is similar to cutting the image like a pie into 36 pieces, with the center of the pie located at the center of the sun. A series of analysis steps are then applied separately to each of the 36 sectors in order to determine the heating power per unit area required by each sector so that it can appear as bright in X-rays as it does in the picture. First, the brightness of the X-rays in the sectors and how this brightness varies with height above the solar surface is compared to a model of the

Sun (Aschwanden and Acton 2001). Based on this comparison, the model outputs values for certain physical quantities that describe the Sun, such as the electron density at the Sun's surface. The numbers obtained from the model can then be used to calculate the rate at which energy is being lost in that sector due to being radiated as light. Since the rate of energy loss must be equal to the rate at which heat energy is supplied, as we described earlier, we now know the rate at which heat energy must be supplied to that sector, F_H. When we have repeated this procedure 36 times, applying it to each sector, we are finally done, since we now have a numerical estimate for the rate at which energy is supplied to each and every sector. This is the power required per m^2, F_H, for each sector.

Goal-Directedness
This last paragraph describes method. The author simply refers to how the data were collected. He continues by describing what the data mean and how information about the rate of energy loss at different regions of the Sun can be extracted from them.

The resulting F_H for each of the 36 sectors, calculated by analyzing the X-ray image, is shown in histogram form in the bottom panel of Fig. 10.1. The power is plotted along the vertical axis as a function of the angular position around the Sun, which is marked along the horizontal axis in degrees from 0° to 360°. In the plot, we mark those sectors that contain active regions with dark gray, quiet-Sun regions with light gray, and coronal holes with white. The heating requirement in Watts per square meter is about $200 \leq P_H \leq 2,000$ W/m^2 in active regions, about $10 \leq P_H \leq 200$ W/m^2 in quiet sun regions, and $5 \leq P_H \leq 10$ W/m^2 in coronal holes, which is in agreement with the estimates of Withbroe and Noyes (1977). If we add up the heating energy requirement in those three categories (active, quiet and coronal holes), we find that the active regions demand 82.4 % of the heating requirement, the quiet-Sun regions 17.2 %, and coronal holes merely 0.4 %. It is clear then, that although active regions do not cover a major fraction of the Sun's surface, the part of the coronal volume that is physically connected to active regions still dominates the total energy requirement of the coronal heating problem. Therefore, we mostly focus our arguments in the following section on active regions, more specifically on coronal loops in active regions and on the heating processes occurring there, rather than on coronal holes.

Argumentative Structure
This last paragraph provides an argument for restricting the bounds of consideration to the active regions on the Sun's surface because they account for over 80 % of the energy requirement. The presumption appears to be that, if a correct account of the source of heating could be given for the active regions, then we would be well along in providing an explanation of the source of energy for the entire Sun's corona.

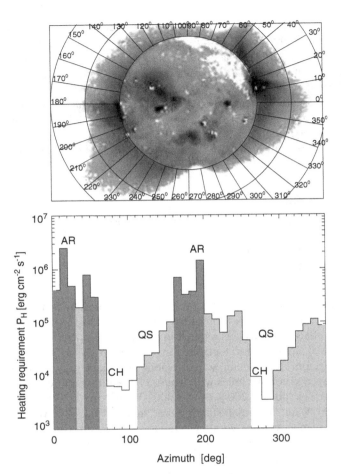

Fig. 10.1 *Top panel*: A soft X-ray image of the Sun observed on 26 August, 1992, with the Soft X-ray Telescope (SXT instrument) onboard the Yohkoh spacecraft. The Sun's corona has been divided into 36 sectors, each 10° wide. The *white circles* indicate distances from the center of the sun of r = 1.0, 1.5 and 2.0 solar radii. The apparent solar surface is marked by the *inner white circle* at r = 1.0 solar radius. Note two active regions at the east and west, a coronal streamer in the southeast and coronal holes in the north and south. *Bottom panel*: The histogram shows the heating rate requirement in the 36 sectors around the Sun. The labels indicate the locations of active regions (AR—*dark gray*), quiet-Sun regions (QS—*light gray*) and coronal holes (CH—*white*) (Aschwanden 2001)

Three Arguments for Heating in the Chromosphere/Transition Region

Temperature Evolution of Coronal Loops

In the previous section, we found that nearly all of the energy required to heat the solar corona comes from active and quiet regions of the Sun. These regions consist mostly of closed magnetic field structures. These structures consist of curved lines of the Sun's magnetic field that rise up from and return to the solar surface, forming giant arches or loops that do not extend far out into space beyond the Sun itself. Most closed magnetic loops have their highest points less than 1 solar radius above the Sun's surface, and are located toward the equator of the Sun. Hot plasma can flow along the closed magnetic field lines, forming coronal loops. In contrast, the polar regions of the Sun are open-field regions, with magnetic field lines that emerge from the Sun and extend far out into space, forming the interplanetary magnetic field. Plasma particles (electrons and ions) can move along open field lines. They are carried far into space beyond the corona, forming the solar wind that flows from the Sun in all directions.

We wish to examine the heating processes that may be occurring along coronal loops in active and quiet regions of the Sun. More specifically, we want to compare two possible heating scenarios. The first scenario proposes that the entire loop of plasma is being heated directly in the corona. We will call this hypothetical process "coronal heating." The second proposal is that the heating first occurs below the corona, in the chromosphere or transition region, and then the hot plasma moves up along the closed magnetic loop into the corona. We will refer to this hypothetical mechanism as "chromospheric heating". For each of the two cases, we will make predictions of what we would observe happening on the Sun if that hypothesis were accurate. Then we will compare observations to our predictions, to see which heating scenario has the most observational support.

Rationality
In the previous paragraph, the author describes the rational side of the fallible rationality epistemology that undergirds science. The author first describes two competing hypotheses (they are competing in the sense that both of them cannot be true) and then lays out how they will be tested. Predictions will be made from each hypothesis, assuming each is true for the nonce, and these predictions will be compared to what has been observed. Predictions that correspond to observations provide reasons in support of the hypothesis. Predictions that do not correspond provide reasons against the hypothesis.

The two competing scenarios—coronal heating and chromospheric heating—are illustrated in Fig. 10.2. First, let us consider the coronal heating scenario, which is shown in the top panels of Fig. 10.2. If magnetic loops are directly heated in the

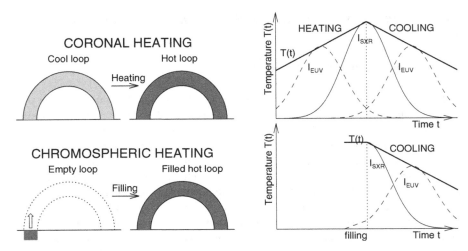

Fig. 10.2 A comparison between the coronal heating scenario (*top panels*) and the chromospheric heating scenario (*bottom panels*). *Top left*: The coronal heating scenario involves the direct heating of a pre-existing cool loop in the corona. *Bottom left*: In contrast, the chromospheric heating scenario fills an empty loop in the corona with hot plasma from below. *Top right*: A plot of temperature versus time shows that in the heating phase of the coronal heating scenario a loop should be observable in EUV light and then in soft X-rays (SXR). As the loop of plasma cools, it will fade in X-rays and become bright again in EUV. *Bottom right*: In the chromospheric heating scenario, the initial brightening in EUV light is missing. Hot plasma from below the corona rapidly fills up the loop and is not directly heated in the corona. Therefore only the cooling phase, from brightness in soft X-rays to brightness in EUV, is expected

solar corona, then the temperature of a coronal loop should increase over some time interval, as long as the heating rate exceeds the rate at which energy is lost by the loop. A loop that begins cool (less than 1 MK) will first appear dark when viewed using an extreme ultraviolet (EUV) filter that allows only EUV light to pass through it and be recorded. This is because at low temperatures the plasma is not hot enough to emit light in the EUV band of wavelengths. As the cold loop is locally heated in the corona, it is expected to first brighten in the EUV filter as the plasma reaches a temperature of between 1 and 2 MK and becomes hot enough to emit EUV light. As the plasma continues to heat, it will appear dim in the EUV filter since it will begin to emit light of a higher energy and shorter wavelength. As the temperature reaches 2–4 MK, the loop should brighten in soft X-ray filters, since it will then begin to radiate X-rays. When the loop starts to go through a cooling phase, the brightening in the light filters should happen in reverse. First the plasma should be bright in the soft X-ray filter and then in the EUV filter. A complete heating and then cooling phase is shown in a plot of plasma temperature vs. time in the top right portion of Fig. 10.2.

Fallibility

In this last paragraph, the author is providing precisely described and timed observations that would be predicted if the coronal heating hypothesis were true. In the following paragraph, the author informs us that the predicted heating in coronal loops (evidenced by brightening in the extreme ultraviolet (EUV) filter has never been observed. As the author reminds us, a failed prediction *might* mean that the hypothesis is false; it might also mean that instrumentation has not been adequate to record the heating event occurring. The outcome is left hanging as in a suspense novel while the author turns to the next hypothesis.

In looking at actual observational data of the Sun, we find that the first part of the coronal loop heating phase (a brightening in EUV) has never been observed, or at least has not been reported in the literature. This could mean one of two things. First, the initial part of the heating phase (EUV brightening) could still be occurring in the corona, but it could be happening very fast—so fast that the instruments, which typically take a measurement every minute, are not able to record it. In this case, our current observations do not contradict the coronal heating scenario. Instead, they are simply not able to show whether it is occurring. The second possible meaning is that the observations show that this heating phase simply does not occur. In other words, the heating of the loop does not occur directly in the corona, and thus the coronal heating scenario is not supported by data.

Let us consider the chromospheric heating scenario, where heating occurs in the upper chromosphere or lower transition region, with subsequent upflows of plasma that fill coronal loops. In this case, we would not detect any heating phase in the coronal part of a loop, and hence would not see an initial brightening of the loop in an EUV filter. This heating scenario has been previously modeled with computer simulations. According to recent simulations (Warren et al. 2002, 2003), as soon as the loop starts to fill with hot plasma, the loop is expected to brighten rapidly in the soft X-ray filter first, during a time interval that is on the order of minutes for typical coronal loops, without any detectable initial brightening in the cooler EUV filters. Therefore, the simulations suggest that in the case of chromospheric heating we would observe only the cooling phase, when the loop cools from the soft X-ray emissions to the EUV light, as shown in the bottom panel of Fig. 10.2. Instances of such loop cooling phases, from the Yohkoh spacecraft's soft X-ray filters and the TRACE spacecraft's EUV filters, have indeed been observed in detail for 11 cases over time intervals of several hours, without any noticeable EUV signature indicating an initial heating phase (Winebarger and Warren 2005; Ugarte-Urra et al. 2006). In a single case (Ugarte-Urra et al. 2006) the footpoint of a loop has been closely monitored in EUV during the initial brightening in soft X-rays of that loop. In this case, the full loop was not detected in EUV ($T = 1.0$–1.5 MK) until it cooled from the hotter ($T = 3$–5 MK) soft X-ray temperatures.

Although observational data seems to support the chromospheric heating scenario, it is not yet possible to conclude that this scenario represents reality. The observational proof is still lacking in a couple of key areas. Firstly, the initial heating agent of the plasma in the chromosphere has not yet been observed. Secondly, the actual upflow of the plasma through the loop has not been observed using Doppler shift techniques. However, this could be due to instrument limitations, such as low sampling rates and poor resolution.

Fallibility
The preceding paragraph displays the massive amounts of caution that scientists can display. Even with the evidence that was sought in hand, the author is still tentative. There remain known considerations that have not been taken into account, so a firm conclusion cannot be drawn.

Overdensity of Coronal Loops

Coronal loops can be detected in images of the Sun only if they have a density that differs from the density of the background plasma that surrounds them in the corona. This is because the brightness of plasma in EUV light and soft X-rays is proportional to the squared density of the plasma. So, features of the corona that are denser than the background plasma will appear brighter than the background, allowing them to be distinguishable in EUV and soft X-ray pictures. In particular, heated coronal loops appear bright in these types of pictures, and therefore coronal loops must be more dense, or 'overdense,' with respect to the background corona. This observed higher density of hot coronal loops is a fact that needs to be explained by every proposed coronal heating theory. If a heating theory cannot account for the observed overdensity of coronal loops, then it must be reworked or discarded.

Argumentative Structure
In the preceding paragraph the author brings into consideration a known fact, that is, that coronal loops are denser than the background corona. In the previous section, the author had considered confirmed or disconfirmed predictions as a source of evidence for or against the hypotheses. Here, the author introduces an already known fact and claims that for a hypothesis to be acceptable it must be able to explain this fact. In the following paragraph, the author argues that between the coronal heating and chromosphere heating hypotheses, only the latter can explain the greater density of coronal loops. It is interesting that the fact about the over-density of coronal loops was known before the time of writing. Perhaps when it was first discovered it was not understood as a fact that could serve as evidence for one hypothesis of why

(continued)

the Sun's corona is hotter than regions closer to the core and against another hypothesis. This case serves to illustrate the points made in Chap. 3 about the fluidity of evidentiary relevance.

Consider theories that involve heating processes in the corona that directly and locally heat coronal loops. In such heating processes (for example, magnetic reconnection; see Aschwanden (2001) for a discussion of theoretical models), the density of the plasma in a coronal loop is not expected to significantly increase. We therefore argue that no theory involving local heating in the corona is able to explain the observed brightness, and thus overdensity, of hot coronal loops. In contrast, the plasma density in coronal loops would be expected to increase significantly due to upflows of additional plasma from the loop footpoints. This would require a heating process that acts below the corona, in the upper chromosphere or lower transition region, with hot, dense plasma moving up into coronal loops (as shown in the bottom of Fig. 10.2). This type of heating theory, which we previously referred to as chromospheric heating, could indeed explain the observed brightness and overdensity of hot coronal loops, and is therefore further supported by EUV and soft X-ray observations.

Upflows in Coronal Loops

The best direct evidence for chromospheric heating, with a subsequent filling of the coronal loop with hot plasma, comes from actual observations of hot upflows from the chromosphere. These plasma flows have been detected by a 'feature tracking' analysis of images of the Sun taken by cameras with special filters onboard the TRACE spacecraft. The upflows are observed to exhibit one-way or two-way flows along loops. An example of such upflows is shown in Fig. 10.3. In this case, the upflow has been inferred from feature tracking in TRACE Fe IX/Fe X (171 Å) images of an active region of the Sun on December 1, 1998 (Winebarger et al. 2001). In the top left panel of Fig. 10.3, a bundle of bright, fan-shaped coronal loops can be seen in the TRACE image. Three specific loops are labeled in the figure. As time passes, later images show an increase in brightness that moves along loop 3. Note, however, that these later images have not been included in Fig. 10.3. Instead, from each of these successive images, a strip of the picture has been extracted along loop 3, and the resulting time sequence of 10 strips that maps the brightness along loop 3 is shown on the right hand side of Fig. 10.3. In this sequence, notice how the peak in brightness moves outward from the loop footpoint, which is located at the left edge of each of the 10 strips.

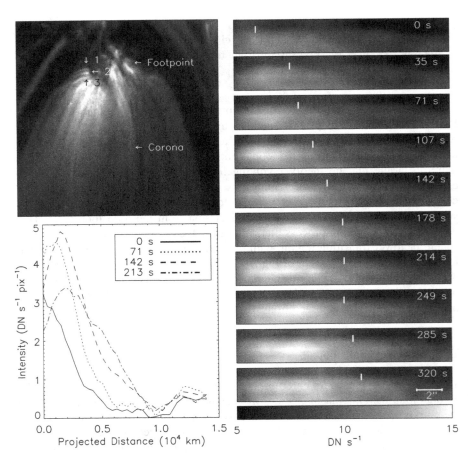

Fig. 10.3 Top left: An image from the TRACE spacecraft in light with wavelengths of about 171 Å (17.1 nm) shows fan-like loops in an active region (labeled AR 8396) observed near the Sun's center on 1 December, 1998, 01:00:14 UT. The field of view of this image is 270 million meters (270 Mm). *Right side*: The brightness along loop 3 over a segment of 15 Mm is shown at 10 time intervals after 01:40:35 UT. The time after 01:40:35 UT is shown in the *top right corner* of each strip. The footpoint of loop 3 is always at the *left side* of each strip. Note how the brightness expands along loop 3, and therefore toward the *right hand side* of each strip, as time passes. The leading edge of the expanding brightness is marked with a *white bar* in each strip (Winebarger et al. 2001)

Fallible Rationality

In the last paragraph, the author declares a crucial aspect of the epistemology of fallible rationality, namely, that the best evidence comes from actual observations. Note, however, what counts as "actual observations of hot upflows from the chromosphere." The upflows are "inferred from feature tracking in TRACE Fe IX/Fe X (171 Å) images". Often, the science curriculum urges students to distinguish observations from inferences, whereas here we read that the author conjoins them. How can what is inferred serve as observational evidence? The caption for Fig. 10.3 reveals what was observed, "the brightness expands along loop 3, and therefore toward the right hand side of each strip, as time passes". The author must presume that this expanding brightness from left to right in the image can only mean that a hot upflow is moving from left to right along a coronal loop. Clearly, it is open for another scientist to deny this presumption and consequently, to deny that hot upflows have been observed moving along coronal loops. In order to defend the presumption, the author would be required to show that no other interpretation of the images is possible.

By following the movement of the intensity peak, it is possible to measure the distance that the peak has moved along the loop and the time it has taken to move this distance. From these quantities, the velocity (speed) of the flow can be estimated. From this series of images, the projected velocity of the plasma flow was measured for four different loops (loops 1–3 and a fourth loop not indicated in the figure). The flows lasted between 2 and 15 min, and the projected velocities were found to be in the range of 5–15 km/s. Note that we can speak only of 'projected' velocities, because the camera can provide only a two-dimensional image. The orientation of each loop in three-dimensional space is not known for sure, and so there are 'projection effects.' This means that the measured velocities give only lower limits to the actual flow speeds.

Models of coronal loops that do not include flows (Winebarger et al. 2001) do not predict the dynamic features (moving intensity peaks) that are observed in this example. This suggests that these dynamic features, in plasmas with a temperature of about 1 MK, represent upflows of heated plasma from the chromosphere into the corona. In modeling work that does incorporate flows, coronal loops are shown to require one-way flow speeds on the order of about 10–40 km/s (Petrie et al. 2003). This speed is consistent with the TRACE observations described above, since we already stated that the measured velocities (5–15 km/s) represent lower limits only. Therefore, heating theories that include upflowing plasma, such as the chromospheric heating theory outlined in this paper, are further supported by feature tracking in TRACE observations.

Argumentative Structure

Although the author has been arguing in favour of the chromospheric heating hypotheses, in the last paragraph it is made clear that the data examined in this section are not sufficient to support that hypothesis over other possible "heating theories that include upflowing plasma". What these data do accomplish is to rule out theories that postulate coronal heating. Support for a hypothesis thus must be interpreted comparatively—although data might be evidence in support of a hypothesis it might support other hypotheses equally well and not be specific enough to either hypothesis to point to one as the most strongly supported.

Conclusions

We have presented three pieces of observational evidence that suggest that the primary heating process leading to the appearance of hot (over 1 MK) coronal loops takes place in the chromosphere, rather than in the corona itself. Specifically, we have pointed out (1) the absence of observed temperature increases in coronal loops, (2) the overdensity of coronal loops with respect to the background coronal plasma, and (3) observed upflows in coronal loops. Given the arguments, we see strong support for the idea that the major heating source responsible for a hot corona is located in the transition region or upper chromosphere, rather than in the corona. We therefore consider the term "coronal heating" to be rather poor, since the essential heating process does not take place in the corona, but rather in the transition region or upper chromosphere. We suggest that the coronal heating problem should instead be referred to as the "chromospheric heating problem," since the existence of the hot corona is a consequence of chromospheric heating and the filling of coronal loops afterwards with hot plasma. This shift of the heating source, from the corona to the chromosphere, rules out models that involve the heating of active region loops directly at coronal altitudes, such as wave heating models. Finally, note that our arguments mostly apply to the heating of active and quiet region loops, but not necessarily to open-field regions, such as coronal holes.

Fallible Rationality

The final paragraph is interesting in its ambiguity. On the one hand, the author speaks of the chromospheric heating as if it is a known fact. On the other hand, we are told that the arguments presented in the paper "suggest that the primary heating process...takes place in the chromosphere", and that there is "strong support for the idea that the major heating source...is located in the...chromosphere". Thus, the evidence suggests or provides strong support but seems insufficient to compel a conclusion. Furthermore, the conclusion is

(continued)

that the major heating source is in the chromosphere leaving open the possibility of some heating taking place in the corona itself. Finally, the conclusions, however tentative or strong, are said to apply only to those areas of the Sun containing "active and quiet region loops". Thus, the work is doubly circumscribed, by qualifications on the strength of the evidence and by limitations to the applicability of the chromospheric heating model. All of this is consistent with the scientific epistemology of fallible rationality.

References

Aschwanden, M. J. (2001). An evaluation of coronal heating models for active regions based on Yohkoh, SOHO, and TRACE observations. *Astrophysical Journal, 560*, 1035–1044.

Aschwanden, M. J. (2004). *Physics of the solar corona: An introduction.* New York: Springer in association with Praxis Publishing.

Aschwanden, M. J., & Acton, L. W. (2001). Temperature tomography of the soft X-ray corona: Measurements of electron densities, temperatures, and differential emission measure distributions above the limb. *The Astrophysical Journal, 550*, 475–492.

Aschwanden, M. J., Winebarger, A., Tsiklauri, D., & Peter, H. (2007). The coronal heating paradox. *Astrophysical Journal, 659*, 1673–1681.

Petrie, G. J. D., Gontikakis, C., Dara, H., Tsinganos, K., & Aschwanden, M. J. (2003). 2D MHD modelling of compressible and heated coronal loops obtained via nonlinear separation of variables and compared to TRACE and SoHO observations. *Astronomy & Astrophysics, 409*, 1065–1083.

Ugarte-Urra, I., Winebarger, A., & Warren, H. P. (2006). An Investigation into the variability of heating in a solar active region. *The Astrophysical Journal, 643*, 1245–1257.

Warren, H. P., Winebarger, A. R., & Hamilton, P. S. (2002). Hydrodynamic modeling of active region loops. *The Astrophysical Journal, 579*, L41.

Warren, H. P., Winebarger, A. R., & Mariska, J. T. (2003). Evolving active region loops observed with the *Transition Region and Coronal Explorer*. II. Time-dependent hydrodynamic simulations. *The Astrophysical Journal, 593*, 1174–1186.

Winebarger, A. R., & Warren, H. P. (2005). Cooling active region loops observed with SXT and TRACE. *The Astrophysical Journal, 626*, 543–550.

Winebarger, A. R., DeLuca, E. E., & Golub, L. (2001). Apparent flows above an active region observed with the *Transition Region and Coronal Explorer*. *The Astrophysical Journal, 553*, L81.

Withbroe, G. L., & Noyes, R. W. (1977). Mass and energy flow in the solar chromosphere and corona. *Annual Review of Astronomy and Astrophysics, 15*, 363–387.

Chapter 11
Maritime Archaic Indians in Newfoundland: An Annotated Example of an Adapted Primary Literature (APL) Article

In this chapter we provide an example of an APL article annotated for epistemology and structure. The APL article originated from a Primary Scientific Literature (PSL) article (Bell and Renouf 2004), proposing an explanation for the puzzling distribution of known dwelling sites for ancient maritime people on the island of Newfoundland. The scientists propose that there is a scattered distribution of living sites along the present coast of Newfoundland. They argue that these sites are scattered due to the fact that the coast has fluctuated dramatically over the last several thousand years because the ocean volume has increased and decreased as the island of Newfoundland has risen and fallen into the Earth's mantle.

The PSL article, which served as the basis for the adaptation, is composed of the following sections: Abstract, Introduction, Archeological context, Relative sea level history, Late Maritime Archaic Indian (MAI) sites and Relative Sea Level (RSL) history, Early MAI sites and RSL history, Early MAI sites on the Northern Peninsula, and Early MAI site preferences, Summary and Conclusion. These sections were partially retained in the APL article which appears below and it is composed of the following sections: Abstract, Introduction, Early and late MAI, RSL in Newfoundland, Late MAI site locations, Interpretations, and Conclusions.

© Springer Science+Business Media Dordrecht 2015
A. Yarden et al., *Adapted Primary Literature*, Innovations in Science
Education and Technology 22, DOI 10.1007/978-94-017-9759-7_11

Did Early Maritime Archaic Indians Ever Live in Newfoundland?[1]

Robert J. Anstey, Trevor Bell and Priscilla Renouf
Memorial University of Newfoundland, St. John's, NL, Canada

Stephen P. Norris
University of Alberta, Edmonton, Canada

Abstract

'Maritime Archaic Indians' is the name archaeologists use to refer to particular groups of prehistoric peoples that lived near the ocean in northeastern North America between about 9,000 years and 3,200 years ago. This paper explores the relationship between relative sea level change and the location of Maritime Archaic Indian habitation sites on the island of Newfoundland.

Keywords

Relative sea level; Maritime Archaic Indian; Newfoundland; Labrador.

Introduction

Relative sea level (RSL) can change over time. RSL refers to the position of sea level relative to the land. RSL can change if the volume of water in the oceans increases or decreases, if the land rises or falls, and if both change at once. Land that is covered by glaciers deforms under the weight of ice and warps downward. As the glaciers melt the land rebounds and, at the same time, the meltwater increases the volume of water in the oceans. The sum of these two processes results in RSL and when one or other dominates, the RSL changes. For instance, if land rebound is greater, then RSL falls; a larger ocean volume change will result in a rise in RSL. As RSL changes, it can affect the shape and position of the coastline. In some regions in the past, RSL changed so much that what once was coastal is now on high

---

[1] This paper is an educational resource produced in 2013 for use in high school science or social studies teaching and adapted from the following original source: Bell and Renouf (2004). Drs. M.A.P. Renouf, an archaeologist, and Trevor Bell, a geographer, both professors at Memorial University of Newfoundland are two of the most prominent scholars on Maritime Archaic Indians.

ground inland from the modern shoreline. In other regions, the ancient coast is now submerged on the modern seafloor.

These changes in coastline have implications for archaeologists who study marine-oriented prehistoric peoples because habitation sites that were once near the coast might now be underwater, eroding into the sea, or indeed well above current sea level. Prehistoric sites of coastal peoples can therefore be quite difficult to find where RSL has changed. Such searches require not only a great deal of luck but also an appropriate method of investigation to determine the former position of the coast. Reconstruction of RSL history has proven to be an important first step for predicting prehistoric site locations.

In this study we examine RSL history and its potential for answering two questions regarding the location and distribution of the earliest prehistoric sites in Newfoundland. The earliest inhabitants of Newfoundland and Labrador are referred to by archaeologists as Maritime Archaic Indians (MAI). Early MAI refers to the earliest groups that lived there before 5,500 years ago. Late MAI refers to groups that lived there between 5,500 and 3,200 years ago. Late MAI sites have been found in both Newfoundland and Labrador. Sites where late MAI lived are unevenly distributed across the island of Newfoundland. No early MAI sites have yet been found in Newfoundland, but are found in nearby mainland Labrador. On these bases we ask: Can RSL history explain (1) the uneven distribution of late MAI sites and (2) the apparent absence of early MAI sites?

The first three paragraphs accomplish several tasks. The first informs readers of the geological processes that might be unknown to them. It is knowledge of these geological processes that helps make sense of the question asked in the title. The second paragraph describes some of the implications these geological processes have for the study of pre-historic coastal peoples. The third points to an anomaly in the archaeological record: Even though Newfoundland and Labrador are adjacent to each other, they appear to have very different histories of MAI occupation. How might this difference be explained? Recognizing an anomalous situation often is key to asking an interesting scientific question such as this one.

Early and Late MAI

One of the ways archaeologists distinguish between early and late MAI is based on a method of evaluating the age of things using knowledge of radioactivity, more specifically knowledge of the radioactive decay of carbon isotopes, which occurs in all organic matter. This method is referred to as radiocarbon dating. Archaeologists also use differences in stone tools to draw conclusions about age. Based on studies of MAI stone tool collections from Newfoundland, Labrador and Quebec, archaeologists have identified two types of stone projectile point: side-notched and stemmed. They link these different shapes to age, with side-notched points dating from 5,500 to 3,200 years ago in southern Labrador and Newfoundland. Stemmed points date from 8,000 to 3,200 years ago in central and southern Labrador. Stone tools from both MAI periods demonstrate the importance of coastal resources for these people.

There has been only a few MAI sites investigated in Newfoundland. Most of these sites have a relatively small amount of remains left from the occupation. There is also currently no evidence of dwelling structures. In Labrador, however, both early and late MAI dwellings have been found. These dwellings can range in size from single-family pit houses to longhouses 12 meters (12 m) to 90 m in length. They can be outlined by gravel or rubble walls, occasionally by sunken floors, and in some cases by a linear pattern of hearths, soil staining and stone debitage. Although the evidence from burial sites suggests that coastal resources were important, the limited evidence of dwellings and the low volume of remains suggest a less intense occupation of the coast by MAI in Newfoundland. This conclusion follows because, if these people were occupying coastal areas on a longer term basis, their habitation sites would have more remains, deposited through generations of use. Likewise, if they were staying at one site for an extended period of time, they would need shelter and presumably evidence of such shelter would have been found. These conclusions suggest that MAI in Newfoundland were fairly mobile, moving to and from resources as they became available and thus making relatively short stays at sites.

Fallibility
In these two paragraphs additional information is provided on the known differences between early and late MAI. The second paragraph points at a possible partial explanation of the anomaly introduced in the previous section. It is possible that the MAI did not occupy Newfoundland as extensively as Labrador despite the proximity of the landmasses. Nevertheless, the authors do not give up on the idea that early MAI may have lived in Newfoundland, even though they may have been more mobile than they were in Labrador. Thus, the authors approach their exploration with a level of openness that allows for multiple possible outcomes.

RSL in Newfoundland

Sea levels fluctuate. One cause of fluctuation is change in ocean volume due to formation and melting of continental ice sheets during and between Ice Ages, referred to as glacio-eustasy. Ocean volume can decrease when ocean water is frozen within ice sheets resulting in global sea level lowering. Ocean volume can increase due to meltwater returning to the oceans from melting ice sheets causing sea level to rise. During the last Ice Age, global sea levels fell and rose by more than 100 m, causing temporary land bridges to form where today there are seas (e.g., Alaska was connected to Siberia across the Bering Strait). Another cause of RSL change is vertical movement of the Earth's crust due to loading and unloading by ice sheets, referred to as glacio-isostasy. As ice sheets melt, their weight decreases. The reduced weight releases pressure on underlying land masses. As pressure is released, landmasses rebound, rapidly at first and then slowly over 10,000 years or so (known as glacio-isostatic rebound). Regions of Canada that were under the thickest part of the last ice sheet are still rebounding today (e.g., Hudson Bay).

Reconstruction of post-glacial RSL history in Newfoundland has relied on the use of sea-level index points from both onshore and offshore locations. An index point is a documented sea-level elevation for a particular feature or location with an established age for when the sea stood at that level. The age of these levels is determined through the radiocarbon dating of associated organic materials, such as marine shell and bone. Index points from a variety of different levels in a particular region can outline the relative sea level change over time in that region.

Since the end of the last glaciation about 12,000–10,000 years ago, RSL has varied significantly in Newfoundland, both through time and across regions. This wide variation contrasts with southern Labrador, which has had a relatively straightforward post-glacial RSL history. In southern Labrador, after the last glaciation RSL fell quite rapidly to near its present level. Geophysicists have generated models that predict the history of RSL fall. Archaeological surveys in southern Labrador support such models, with early MAI sites occurring at higher elevations and farther from the coast than late MAI sites. In Newfoundland, however, RSL was complicated by the combined effects of being near the margin of a continental ice sheet and having local ice centers. Taking these factors into consideration, one geophysical model published for Atlantic Canada in 1981 by geophysicists Gary Quinlan and Christopher Beaumont predicted three types of RSL history for different regions of Newfoundland. Type A predicts continuous RSL fall in the northwest and hence all ancient shorelines should be preserved above the modern coast. Type B, relevant for most of the island, predicts an initial RSL fall to a sea-level lowstand, followed by an RSL rise to present. Under the Type B scenario, the age and depth of the lowstand is an indication of how much of the RSL record is submerged on the seafloor in contrast to raised on the landscape. An older deeper lowstand means a significant portion of the RSL record is now submerged, while a younger, shallower lowstand implies a substantial RSL record above present sea level. Type C predicts RSL changes entirely below modern sea level for eastern-most Newfoundland. The data from sea-level index points generally support this three-part model.

Argumentative Structure

The previous paragraph begins to explain why the history of MAI may appear differently in Labrador than in Newfoundland. The regions had RLS histories radically different from each other. It is this difference in RSL history that the authors hope to exploit in explaining another difference, that of different patterns of evidence of MAI habitation in Newfoundland and Labrador.

As shown in Fig. 11.1, the depth of the post-glacial lowstand in Newfoundland generally decreases towards the heads of the major bays and northwest towards the Northern Peninsula. Where the lowstand was deeper than 25 m below present sea

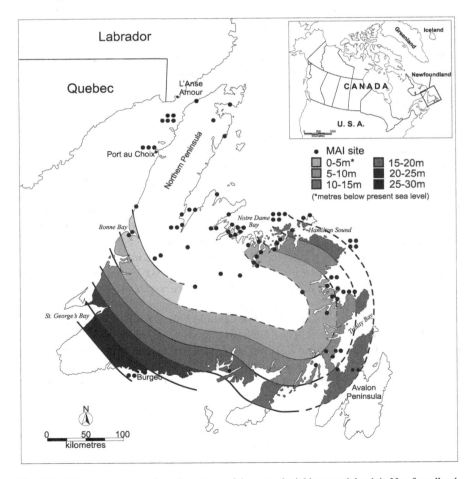

Fig. 11.1 The map contours show the pattern of the post-glacial lowstand depth in Newfoundland. The *dots* show the locations of late MAI sites. In areas where the lowstand was deeper than 15–20 m, late MAI sites were probably submerged by rising RSL

level it was probably earlier than 9,500 years ago and where it was shallower than 16 m below present sea level it was probably later than 8,500 years ago. In areas where the lowstand is deeper than 15–20 m, late MAI sites would probably be submerged.

Fallibility

Note the uncertainty expressed in the previous paragraph. The word "probably" is used to modify three estimates, two dealing with timeframes and one of whether late MAI sites would now be submerged in areas where the postglacial lowstand is deeper than 15–20 m. Several uncertainties come together: the exact timing of the end of the last glacial period; the timing of possible late MAI habitation; and the rate of rise of RSL. Therefore the authors judge that they are warranted only in a conjecture that some late MAI sites would be submerged at the present time.

Late MAI Site Locations

To date, late MAI sites have been found in both Labrador and Newfoundland. Whereas early MAI sites have been found only in mainland Labrador, for the present study, we examined the location characteristics of 80 late MAI sites in Newfoundland. We found that 84 % of these sites are located along the coast and are concentrated in areas where marine food resources (e.g., seals, birds, fish and shellfish) were relatively abundant and close to shore. Sixteen percent are located in the interior. At a smaller spatial scale, we found that most sites are in nearshore (71 %) rather than offshore (13 %) locations. Based on an examination of coastal sites only (n = 67), we found that most (63 %) are located in sheltered coves. Sites are regularly found near a river, stream or pond. They also have a view in more than one direction and are near heights of land, which could be used for lookouts. A comparison of known site elevations with the RSL record tells us that most late MAI coastal sites were within 5 m of the shore during their respective occupations. We examined the distribution of late MAI sites and found that the highest concentration (61 %) of known sites is in northeast Newfoundland between Notre Dame Bay and Trinity Bay. There are no known late MAI sites on the west coast south of Bonne Bay and very few on the south coast or Avalon Peninsula.

Argumentative Structure

In the preceding paragraph, the authors adopt a tone of greater certainty than that expressed in the previous section. This is so, because the authors are describing what they take to be known about the late MAI site locations as opposed to forming hypotheses about the location of not yet discovered sites.

Interpretations

Late MAI Sites and RSL History

If we return to Fig. 11.1 we can see that the gap in late MAI site distribution coincides with the region of maximum lowstand depth. There are no known MAI sites in St. George's Bay on the southwest coast, where the lowstand reached 25 m below present sea level and 10 m of marine submergence has occurred since late MAI occupation of Newfoundland. In the Burgeo region, on the south coast, the post-glacial lowstand reached 27 m below present sea level about 600 years earlier than St. George's Bay. Since 5,500 years ago RSL in the Burgeo region has risen 8 m above its former level. Only three MAI sites have been identified in this region. These sites are between 7 and 12 m above where the sea level would have been during their occupation. This suggests that they were once on high land. Today these sites are exposed only at low tide. In contrast, if we look at the Hamilton Sound region along the northeast coast where MAI sites are more common, the lowstand reached 17 m below present sea level and RSL has risen only 1–3 m since late MAI occupation.

> **Argumentative Structure**
> The authors have begun to present their interpretations of the data. In the first of the two paragraphs the authors amass the facts they wish to draw upon. In the second paragraph, they explicitly signal their reasoning based upon the facts through the use of the word "infer". It can be assumed from the context that the subsequent sentences are also the result of inferences. The authors thus explicitly differentiate between the facts that they hold relatively firmly and their conclusions inferred based upon these facts, which they hold relatively less firmly.

Based on these data, we infer that the uneven distribution of late MAI sites in Newfoundland is linked to RSL changes after that occupation. Sites along the coast that have experienced more than 5 m of marine submergence since 5,500–3,200 years ago are today submerged. The exceptions are sites on higher elevations (greater than 7 m above sea level) at the time of occupation, which remain at or above present sea level. Late MAI sites that are preserved above sea level occur in regions that have had very little or no marine submergence in the last 5,000 years.

> **Rationality**
> The reasoning has shifted to focus on locations of early MAI sites. Two conclusions are drawn. The first conclusion applies to most of Newfoundland, which the authors point out, experimental lowstands between 8,000 and

(continued)

6,000 years ago that were more than 10 m below their present level. Given that this RSL history is qualified as "likely", it must be assumed that the conclusion that early MAI sites for most of Newfoundland are "submerged today" itself is no more certain than "likely". The conclusion regarding early MAI sites on the Northern Peninsula seems to be presented as more certain. The RSL history for that region is less complex and therefore easier to be confident about: namely, the Northern Peninsula never experienced marine submergence. If this is true, then early MAI sites would not have been eroded by the sea and thus more likely to be located there than elsewhere on the island.

Early MAI Sites and RSL History

In order to assess whether the absence of early MAI sites is related to RSL history we must examine the RSL record for early MAI occupation. Figure 11.2 shows where RSL was likely positioned relative to the present sea level at 6,000 years, 7,000 years and 8,000 years ago. For most of Newfoundland, with the exception of the Northern Peninsula and western Notre Dame Bay, RSL was below its present level between 8,000 and 6,000 years ago, as indicated by the negative numbers. Most of these regions experienced RSL lowstands more than 10 m below their present level. This was of sufficient magnitude to cause early MAI sites to be submerged today. As we mentioned earlier, the Northern Peninsula is the only region with a Type A sea-level history. It thus never experienced any marine

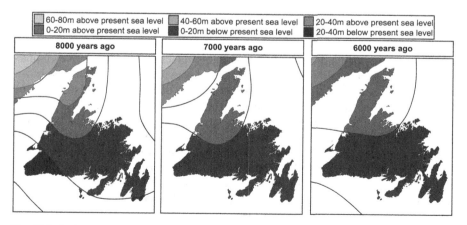

Fig. 11.2 Isobase maps for 8,000, 7,000 and 6,000 years ago modified from Shaw et al. (2002). Individual isobases were interpolated from point data derived from local RSL curves. Where the isobase value equals zero it indicates that the modern shoreline intersects the ancient shoreline

submergence and ancient shorelines are found inland from the present coast. Therefore MAI sites that were once close to the shore are now farther inland. Since all sites on the Northern Peninsula are above water and have likely not been eroded, this region has the best potential for finding early MAI sites.

Most known late MAI sites on the Northern Peninsula are found at 6–10 m above present sea level. They are set back 50–150 m from the present-day shore. In addition, most are located near rivers or ponds. Considering these late MAI site locations, early MAI sites must be at higher elevations and therefore even further back from the present-day shore. Extrapolating from these patterns to early MAI site locations, we propose that early MAI sites on the Northern Peninsula are most likely to be found at elevations greater than 14 m above present sea level. In addition, based on the information on late MAI site settings, these sites are also likely to be found in areas of high marine productivity, at nearshore rather than offshore locations, in a sheltered area such as a cove, near a river, stream or pond, with a view in more than one direction, and near spots of high land. Due in part to the thick layer of peat as well as dense vegetation that covers much of the Northern Peninsula, many of these locations are today difficult to access or identify. This may explain the absence of known early MAI sites on the Northern Peninsula.

Fallibility
The authors assemble many of the facts presented to this point and argue for a proposal about where early MAI sites are "most likely to be found" on the Northern Peninsula. Thus, although they present a strong case, the authors recognize with explicit language the fallibility of their conclusions.

Conclusions

Can RSL history explain the uneven distribution of late MAI sites in Newfoundland? Through examination of the RSL history of Newfoundland, which was shown to be variable and complex, we were able to conclude that the uneven distribution of late MAI sites was likely related to differing RSL around the island of Newfoundland. Late MAI sites that are preserved above sea level occur in regions that have had very little or no marine submergence in the last 5,000 years. Our second question asked whether RSL history could explain the apparent absence of early MAI sites in Newfoundland. We examined the RSL history of the early MAI period and concluded that early MAI sites, if they exist, are likely submerged in most regions of the island. The one exception is the Northern Peninsula, which has not seen any marine submergence since the early MAI period.

Fallible Rationality
The authors return to their two research questions and conclude that RSL history is likely related to the uneven distribution of late MAI sites and to the as-yet failure to discover any early MAI sites. Once more, they qualify their reasoning for conclusions using expressions and words such as "likely", "apparent", and "if they exist".

References

Bell, T., & Renouf, M. A. P. (2004). Prehistoric cultures, reconstructed coasts: Maritime Archaic Indian site distribution in Newfoundland. *World Archaeology, 35*(3), 350–370.

Quinlan, G., & Beaumont, C. (1981). A comparison of observed and theoretical postglacial relative sea level in Atlantic Canada. *Canadian Journal of Earth Sciences, 18*, 1146–1163.

Shaw, J., Gareau, P., & Courtney, R. C. (2002). Paleogeography of Atlantic Canada 13–0 kyr. *Quaternary Science Reviews, 21*, 1861–1878.

Epilogue: The Basis of and Possibilities for Adapted Primary Literature

If it be true that good wine needs no bush, tis true that a good play needs no epilogue; yet to good wine they do use good bushes, and good plays prove the better by the help of good epilogues. (Shakespeare, *As You Like It*, Rosalind's Epilogue)

The above quotation from Shakespeare offers an opportunity to have the last word before we leave the stage on which this book was set. We are excited by the possibilities for inquiry through the processes of the reading, the writing, and the talking about authentic scientific articles through the use of an innovative educational text genre Adapted Primary Literature (APL). The context for the use of APL was motivated by our claim that the language of science is not the problem with students' reading. We understand that high school students cannot read primary scientific articles well but as we describe in this book we developed a means to adapt the primary literature to the knowledge level of the students thereby making it possible for them to read authentic scientific text. Many questions will remain, however this foundational book on APL provides a heightened range of insight that will engage, clarify and provoke the consideration of alternative approaches for the effective teaching of science.

APL provides a means to bring authentic scientific text into the classroom to complement currently used materials and an opportunity to incorporate authentic science practices into the teaching and learning of science. Learning from scientific texts is undoubtedly challenging because of the inclusion of charts, diagrams, tables, images, symbols, and formulas. Moreover, scientific texts are challenging due to their unique structure, epistemology, topic-specific vocabulary, and include meta-scientific language. Scientific texts present a new mode of information presentation and argument over non-scientific texts and this code has to be deciphered by students if they are to become effective readers of authentic scientific materials. APL provides a unique means of science communication through the adaptation of authentic content to the comprehension level of the students.

The case was made over two decades ago that science curricula frequently does not take into account the role of practical reasoning in the production of scientific knowledge (Norris 1992). APL may now fill that void through its focus on practical

© Springer Science+Business Media Dordrecht 2015
A. Yarden et al., *Adapted Primary Literature*, Innovations in Science
Education and Technology 22, DOI 10.1007/978-94-017-9759-7

reasoning. APL as a genre includes how objectivity is established through a redacted style of any reference to primary scientific literature (PSL). The removal is deemed appropriate for K-12 students because it is highly unlikely that they will pursue further reading on a topic. Readers' comprehension of text genres and in particular the expression of uncertainty through hedging in scientific text is inimitable in PSL. Uncertainty is manifested through hedged propositions that offer readers an opportunity to sense the degree of uncertainty in what is written while affording their involvement in assessing, modifying and refining scientific information. This method is a marked departure from the more prevalent portrayal of scientific statements as statements of fact seen in many textbooks. The exploration of scientific language is made possible by APL through adherence to the practices of science including its argumentative structure, scientific content, organizational structure, and uncertainty. These practices situate the knowledge of science in authentic contexts that are relevant in any dialogue on reading in science. Thus, high school students who study science using APL, for example, have the opportunity to study the canonical characteristics of PSL retained in the APL genre for an authentic science experience in how and why science is communicated.

The communication of science is often to promulgate new knowledge that either confirms or disconfirms extant theories or facts. The science found in school texts is radically different from authentic science. APL provides a means to bridge the gap between the community of learners and the community of scientists through understanding the differences in how members of each community think and deal with fundamentals such as evidence, uncertainty, and argumentation. This initiation into the ways of knowing engages students in an authentic scientific experience and enhances scientific literacy in both the fundamental (being able to read, interpret and write scientific text) and derived (the knowledge of scientific ideas and the ability to use them in a scientific manner) senses (Norris and Phillips 2003). The fundamental and derived aspects of scientific literacy are symbiotic and the ability to think scientifically is one aspect of the derived sense.

Scientific literacy and inquiry are both cognitive processes. Inquiry, from an epistemological perspective, is the doing of science (Schwab 1962) of which interpretation, being knowledgeable, and engaging in the learning and doing of inquiry thinking (Tamir 1985) through the use of authentic scientific texts would bridge the gap with school science.

New Directions

It is likely the case that the promotion of inquiry in school science has been longstanding. Research on the nature of the texts used and how best to use them has also been longstanding. The empirical basis of science in curricula has been studied thoroughly, however the basis of science in literacy is relatively new (see a special issue of the *Journal of Research in Science Teaching (1994, Volume 31, Issue 9)*. We make the case that reading is important to science and to students. This is a new direction for school science teaching and learning that can be realized with APL. Through the three meanings of structure (organizational, goal-directed, and argumentative) in science texts we have shown in Chap. 3 that there is considerable variability both within and between science fields. The implication is that flexibility in one's conception of text structure is a worthy stance.

The representation of epistemology in scientific texts evident in the argumentative structure is the signification of reasons, data and evidence for conclusions coupled with the expression of tentativeness in the findings. These define the epistemological foundation in scientific texts and express tentativeness (fallible rationality) in the findings of the five papers compared and contrasted (see Chap. 3 for details). Tentativeness is noted through the extensive use of specific qualifiers, an aspect of meta-scientific language which allows speakers, readers, and writers to refer to scientific practices as well as talk about them in theoretical, critical and pragmatic ways in order to make an evaluation of whether the evidence should be accepted or rejected. Learning how to comprehend and tolerate the tension inherent in the epistemology of science provides a window into how scientists think and marks a major accomplishment for APL that is virtually non-existent in science education where many texts treat science as a black and white process.

It is through the reading of scientific articles that students acquire an understanding of how scientists think. These articles are "authentic specimens of enquiry" (Schwab 1962, p. 81) because they are a record of scientific reasoning, a record of the inquiry process. This record is an entrance to the source of and grounding for knowledge which is the first step to teaching students to critically appraise that knowledge in order to develop self-fulfillment and autonomy. APL maintains the canonical form of the PSL and inspires and supports inquiry in the teaching of school science (Baram-Tsabari and Yarden 2005; Norris et al. 2012). Students noted the difference in the APL articles from their school texts. There was less listing of facts to be remembered and more arguments based on evidence to engage and support students to think, offer challenges and entertain options. This modeling has the potential to be invigorating for students because they can offer a critical response to text that goes well beyond rote responses.

We have worked through the various ways that inquiry is represented in APL through the representation and significance of conceptual understanding of science as well as the nature of science (including tentativeness and argumentation). The outcome of APL and reading in science classrooms has been detailed (see Chap. 4). We argued that reading authentic scientific text is a desirable goal of science education because reading is an essential scientific practice, an essential goal of scientific literacy, must be critical, and provide insight into the nature of science. The implication of these claims is that reading authentic scientific text is the foundation for the most important goal of science education, critical scientific literacy. APL is an impressive evidence-based tool, founded on authentic specimens of inquiry, for the teaching of reading in science.

For the first time, we have provided a step-by-step description of how to create and use APL (see Chap. 5). A suggested sequence of steps are described from choosing a suitable article for adaptation to implementing the article(s) in classrooms. The instructional approaches developed for the use of APL include conversational, problem solving, and scientific literacy. The benefits and limitations of the instructional approaches are discussed and followed by ways to support teachers in the use of APL through workshops, a teachers' guide with video teaching episodes, and numerous questions and activities to accompany each section of the APL implementation process. Research (Falk et al. 2005) showed the videotaped episodes facilitate a reproduction of teaching that occurs while working through an APL article and teachers reported the question pool to be an adaptable teaching tool. Students are expected to read the APL if it is to be a useful and important component of science education.

The nuances of APL applications and their concomitant outcomes are informative (see Chap. 7 for details). The laboratory setting (without teacher intervention) revealed a complexity of reading an APL article not addressed to this point. Two twelfth grade students were asked to read an APL article in developmental biology and they could not discern distinct studies when the same methods were used to answer different research questions (Brill et al. 2004). Cognitive overload was more evident when students were to integrate information pieces from different locations throughout the text (global level) than when the information was from the same location (local level). In science classrooms with teacher intervention, the APL articles prompted more engagement, integration of information, inquiry as well as comprehension of discipline-specific subject matter. Students, however, were limited in knowing when and how to use their prior knowledge to read and interpret well (Falk et al. 2008).

The outcomes of combining APL with other genres (see Chap. 2): Primary Science Literature (PSL), Journalistic Reported Versions (JRV, or popular articles) in the two aforementioned contexts did not lead to different learning outcomes even though each held distinctive genre features (Baram-Tsabari and Yarden 2005; Norris et al. 2012). For instance, the students performed equally well with summarizing the main ideas of the articles. However, those reading the APL displayed a more comprehensive understanding of the processes and methodology of scientific inquiry through asking more questions for future investigations, thinking critically and offering future-application ideas, than those reading the JRV article. These cognitive abilities are evidenced at all levels of schooling.

The basis for a learning progression is emerging from the research completed thus far including that used in sixth grade where authentic APL texts are integrated with narrative writing as Hybrid Adapted Primary Literature (HAPL) (Shanahan 2012; Shanahan et al. 2009) as well as APL in eighth grade (Ariely and Yarden 2013). These studies and others signal that the application of APL is not limited only to high school. Consequently, APL affords exciting opportunities for students to acquire familiarity with the structure of scientific text early in their schooling thereby likely enhancing the sophistication of future high school students.

We provide a compendium of four key APL articles annotated for particular and specific instructional aims (1) *Developing an Inhibitor of Anthrax Toxin* annotated for epistemology, (2) *An Epidemiological Model for West Nile Virus: Invasion*

Analysis and Control Applications annotated for structure, (3) *The Coronal Heating Problem* annotated for epistemology and meta-scientific language; and, (4) *Did Maritime Archaic Indians Ever Live in Newfoundland?* annotated for epistemology. Taken together this set of papers represent communal social practices (see Chap. 3) that produce science. At the core of these practices within each paper is an argumentative structure, and a degree of tentativeness—fallible rationality— where conclusions must conform to the evidence provided at the time but which may change as new evidence is established over time.

In each case, the authors of these articles make an evaluation based on relevant evidence. The point of the annotated articles is to highlight the importance of attending to and monitoring scientists' meta-scientific language, to experience firsthand that science has a literacy in addition to an empirical basis and must thus be interpreted. In addition, the structure (organizational, goal-directed, and argumentative) also joins text and thus more than the words of the text are crucial to comprehension and interpretation. Moreover, students must learn to accept the tension that scientists experience as they endeavor to make the strongest case possible on the basis of the evidence to hand and at the same time address that they may be mistaken.

A Call for Action

It is expected that students will require specific instruction on how to read in a scientific context such as APL as is also the case with most texts. Based on a large body of research (see Chap. 6), it is incumbent on us to conclude that students are weak in their critical reading of science. It is clear that there is a profound need to change what it means to read science. It is a change necessary for K to 12 students with the focus on steadily increasing levels of reading sophistication. Science reading instruction requires teachers who are comfortable with science, who under- stand its basic epistemology, and accept that science is dynamic. Reading in science must be taught in high school through an emphasis on text structure, text episte- mology and meta-scientific language (see Chap. 3). We provide a series of take- home messages (T-hM) that shape and restore a vigor and freshness to pre-service and in-service teacher education on reading in science. At a general level they include:

1. Reading programs provide no reliable direction to teachers on which selections are science;
2. How teachers perceive and understand the text affects how they teach reading in science;
3. Children's perceptions of science are affected by the content focus chosen by the teacher of science reading;
4. The meta-language that structures scientific writing and reveals scientific rea- soning should be an object of instruction in reading science; and

5. Every K-8 teacher and every science teacher should be a teacher of science reading (see Chap. 6).

We have reflected on the contents of the chapters as a starting point to harvest ideas for next steps in the advancement of APL including the formal adoption of APL in the teaching and learning of science education in more jurisdictions around the world, the development of readily available APL materials for various grade levels, the use of APL articles as a model for scientific writing of inquiry projects carried out in schools, the promotion of critical thinking among school students through the use of APL articles that present conflicting or contradictory claims, and the stimulation of the quality and quantity of research on the effective uses of APL in student science learning. In closing, we do not wish to make the claim that our epilogue is good as suggested by Shakespeare but we intend it to be helpful.

References

Ariely, M., & Yarden, A. (2013). Exploring reproductive systems. In B. Eylon, A. Yarden, & Z. Scherz (Eds.), *Exploring life systems (Grade 8)* (Vol. 2). Rehovot: Department of Science Teaching, Weizmann Institute of Science.

Baram-Tsabari, A., & Yarden, A. (2005). Text genre as a factor in the formation of scientific literacy. *Journal of Research in Science Teaching, 42*(4), 403–428.

Brill, G., Falk, H., & Yarden, A. (2004). The learning processes of two high-school biology students when reading primary literature. *International Journal of Science Education, 26*(4), 497–512.

Falk, H., Brill, G., & Yarden, A. (2005). Scaffolding learning through research articles by a multimedia curriculum guide. In M. Ergazaki, J. Lewis, & V. Zogza (Eds.), *Proceedings of the Vth conference of the European researchers in didactics of biology (ERIDOB)* (pp. 175–192). Patra: The University of Patras.

Falk, H., Brill, G., & Yarden, A. (2008). Teaching a biotechnology curriculum based on adapted primary literature. *International Journal of Science Education, 30*(14), 1841–1866.

Norris, S. P. (1992). Practical reasoning in the production of scientific knowledge. In R. A. Duschl & R. J. Hamilton (Eds.), *Philosophy of science, cognitive psychology, and educational theory and practice* (pp. 195–225). Albany: State University of New York Press.

Norris, S. P., & Phillips, L. M. (2003). How literacy in its fundamental sense is central to scientific literacy. *Science Education, 87*, 224–240.

Norris, S. P., Stelnicki, N., & de Vries, G. (2012). Teaching mathematical biology in high school using adapted primary literature. *Research in Science Education, 42*, 633–649.

Schwab, J. J. (1962). The teaching of science as enquiry. In J. J. Schwab & P. F. Brandwein (Eds.), *The teaching of science*. Cambridge: Harvard University Press.

Shanahan, M. C., Santos, J. S. D., & Morrow, R. (2009). Hybrid adapted primary literature: A strategy to support elementary students in reading about scientific inquiry. *Alberta Science Education Journal, 40*(1), 20–26.

Shanahan, M. C. (2012). Reading for evidence in hybrid adapted primary literature. In S. P. Norris (Ed.), *Reading for evidence and interpreting visualizations in mathematics and science education* (pp. 41–63). Rotterdam: Sense Publishers.

Tamir, P. (1985). Content analysis focusing on inquiry. *Journal of Curriculum Studies, 17*(1), 87–94.

Appendices

Appendix A: An Example of an Adapted Primary Literature (APL) Article

Absence of Skeletal Muscle in Mouse Embryos Carrying a Mutation in the Myogenin Gene[1]

Anat Yarden and Gilat Brill
Department of Science Teaching, Weizmann Institute of Science, Rehovot, Israel

Abstract

The *myogenin* gene encodes a transcription factor that is specifically expressed in muscle cells. To investigate the role of the *myogenin* protein in the whole organism, mutant mice were created in which the active *myogenin* protein is absent. The mutant embryos completed their embryonic development in their mother's womb, but the offspring died immediately after birth. They were found to lack skeletal muscle. These results suggest that the *myogenin* gene is essential for the development of skeletal muscle during embryonic development.

Introduction

Myogenin belongs to a family of proteins that function in the cell as **transcription factors**. These transcription factors bind specific DNA sequences and cause

---

[1] This paper is an adaptation based upon Hasty, P., Bradley, A., Morris, J. H., Edmondson, D. G., Venuti, J. M., Olson, E. N., & Klein, W. H. (1993). Muscle deficiency and neonatal death in mice with a targeted mutation in the *myogenin* gene. *Nature, 364*, 501–506.

© Springer Science+Business Media Dordrecht 2015
A. Yarden et al., *Adapted Primary Literature*, Innovations in Science Education and Technology 22, DOI 10.1007/978-94-017-9759-7

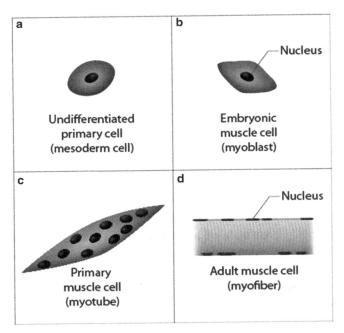

Fig. 1 Schematic illustration of the differentiation process of an adult muscle cell. The differentiation process starts from an undifferentiated primary cell (**a**) from the mesoderm that differentiates into embryonic muscle cell (myoblast); (**b**) The shape of the myoblast in culture resembles a star. Further on in the differentiation process, some myoblasts become confluent, lying alongside each other, and fuse to create a primary muscle cell (myotube); (**c**) A myotube is a multinucleate cell (syncytium). High levels of actin and myosin proteins are synthesized in the myotube, and the cell transforms into muscle fiber (myofiber); (**d**) Along the myofiber, actin and myosin fibers, which are responsible for fiber contraction, are observed

differentiation of muscle cells. Therefore, the genes encoding these transcription factors are called myogenic genes (genes producing muscle, in Greek myo = muscle, genesis = creation). In addition to the gene encoding *myogenin*, the following genes belong to the myogenic gene family: *MRF4*, *Myf5*, *MyoD*.

In an experiment performed on **cell cultures**, each of the myogenic genes was inserted separately by genetic engineering methods into cells that were not muscle cells. In cells in which a myogenic gene was inserted, it was translated into protein. As a result, these cells differentiated into muscle cells, similar to the differentiation process of muscle cells that normally occurs in the whole organism (Fig. 1a–d): first, they resembled embryonic muscle cells (myoblasts, in Greek myo = muscle, blast = bud, Fig. 1b); then they expressed muscle-specific genes, and fused to become multinucleate muscle cells (myotubes, Fig. 1c); finally, they created adult muscle fiber (myofiber, Fig. 1d), and even showed contractions in culture.

Despite the fact that many studies have concentrated on the roles of myogenic genes in experiments performed in cell culture, little is known about their role during embryonic development. It has been found that the four myogenic genes are not expressed at the same time in the mouse embryo: first, the *Myf5* gene is expressed, then, respectively, the genes *MyoD*, *myogenin* and *MRF4*. This difference in the timing of the expression of each of the myogenic genes led to the hypothesis that each of these genes has a different role during development.

To determine the role of a certain gene during embryonic development, it is possible to add the gene, in culture or in a whole organism, into cells that do not normally express it and investigate what changes after its addition. This approach is called **gain of function**. Another approach is to harm a gene by a mutation that renders the protein produced by the gene inactive. When a mutation occurs in both copies (or alleles) of the gene, no active protein is produced in the cell. This approach is called **loss of function**.

To determine the role of each myogenic gene during development, mice were created in which one of the myogenic genes was inactive. Mouse embryos without *Myf5* developed normally and had normal muscles. A similar result was obtained with embryonic mice lacking *MyoD*. These results led to the conclusion that neither of these genes alone is essential to the normal process of skeletal muscle differentiation.

In this study, we investigated the influence of the absence of *myogenin* on mouse embryonic development. For this purpose, mice lacking *myogenin* were prepared. The embryonic developmental process of these mice was completed but the embryos were born without the ability to move and died immediately after birth. The embryos were found to lack differentiated skeletal muscle throughout their bodies. These findings show that *myogenin* is essential for controlling normal differentiation of the skeletal muscles in embryonic development.

Materials and Methods

Creating Mice Without *Myogenin*

Using molecular methods, a mutation in the *myogenin* gene is created such that the resulting mutant gene encodes inactive *myogenin* protein. These mutant genes are inserted into special cells growing in culture (Fig. 2). These cells, called **embryonic stem cells**, are taken from mouse embryos at early stages in their development. Their developmental potential is high, namely, under suitable conditions, most cell types making up the body of the mouse can develop from them. The *myogenin* mutant genes are engineered such that when such a mutant gene penetrates the cell, it can integrate into the genome of the cell and replace one of the two copies (alleles) of the *myogenin* gene. Not all cells incorporate the mutant gene. However, after strict selection processes (which are not described here), only cells in which the mutant *myogenin* gene has integrated survive in the culture. These cells are injected into normal mouse embryos at the blastula stage (Fig. 2) and integrate into

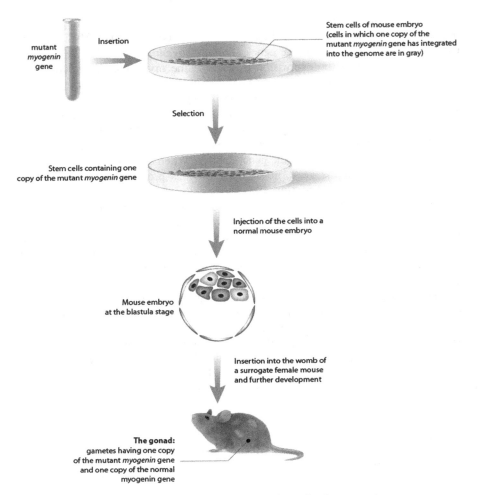

Fig. 2 Diagram of the process used to create mutant embryos for the *myogenin* gene

the cells that already exist in the embryo. The injected embryos are further grown under conditions allowing normal development.

The mice that develop after these injections are **chimeric**: some of their body cells have two normal copies of the *myogenin* gene, and some have one normal *myogenin* gene copy and one mutant *myogenin* gene copy (Fig. 2). If some cells carrying one mutant copy of the *myogenin* gene differentiate into reproductive cells in the chimeric mice, the mutant gene will be inherited by their offspring.

After repeated cross-breeding between the chimeric mice, and later on between their offspring, three types of mice are obtained: (1) mice whose body cells have two copies of the normal gene (homozygous for the normal *myogenin* gene); (2) mice whose body cells have one copy of the mutant gene and one copy of the normal *myogenin* gene; (3) mice whose body cells have two copies of the mutant gene (homozygous for the mutant *myogenin* gene). These latter mice lack active *myogenin*. The basis for the experiments described in this article are the mice without *myogenin*, and the mice with one mutant and one normal gene.

Staining

Thin embryo sections were exposed to a color solution for observation under a light microscope, because tissues are usually transparent. The skeletons of the mouse offspring were exposed to a specific color solution that stains cartilage and bone.

Results

Absence of Active *Myogenin* Protein Is Lethal

Embryos having one copy of the mutant *myogenin* gene developed normally. In other words, one copy of the *myogenin* gene is sufficient for normal development. Half the quantity of *myogenin* protein was found in these embryos compared to normal mice. On the other hand, embryos having two copies of the mutant *myogenin* gene were unable to move and died a short time after birth. No *myogenin* protein was found in these embryos.

Traits of Mice Without *Myogenin*

Mice having two copies of the mutant *myogenin* gene were bluish. The embryo's heart was beating at birth and its lungs were normal, but it was unable to move and did not react when touched.

There was a significant difference between the skeletal muscles of embryos with two copies of the mutant *myogenin* gene and normal mice or mice with one copy of the mutant gene. The quantity of muscle tissue was significantly smaller in the embryos with two copies of the mutant *myogenin* gene and their diaphragm muscle was very thin. These embryos' weight was some 15–20 % lower than the weight of normal embryos from the same litter, probably due to the absence of muscle.

The skeleton of embryos having two copies of the mutant *myogenin* gene revealed additional defects in comparison to normal embryos or embryos having one copy of the mutant gene. Particularly notable in the embryos with two copies of the mutant *myogenin* gene was abnormal bending of the spine and defective structure of the ribs compared to normal embryos (Fig. 3). The ribs were twisted,

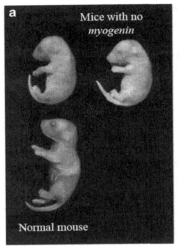  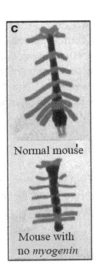

Fig. 3 Comparison of normal mouse embryos to mice without *myogenin*. (**a**) The mouse embryos photographed after birth; (**b**) The spinal cord of the mice; (**c**) The breastbone (in *black*) and the ribs (in *gray*) of a normal embryo and an embryo without *myogenin*, after specific staining of the breastbone and cartilage tissue

some 15 % shorter than in normal embryos, and arranged at an irregular angle relative to the vertebrae of the spinal cord and to the breastbone (Fig. 3). The breastbone was also found to be 50 % shorter than the average length in normal embryos. In general, the skeleton of mice having two copies of the mutant gene was more brittle—probably due to the decrease in muscle mass attached to the bones.

Structure and Composition of Cells and Tissues in Embryos Without *Myogenin*

Observation of tissue sections under a light microscope showed that in places where normal mice have multinucleate muscle cells—such as in the rib area (Fig. 4a), mice without *myogenin* have mononuclear cells and only rarely, muscle fiber (Fig. 4b). A significant decrease in the quantity of muscle tissue was observed in the tongue, in the face muscles and in the diaphragm (the results are not described here).

To examine whether any differentiation was initiated in the muscle cells of embryos that have two copies of the mutant *myogenin* gene, the quantity of muscle-specific proteins (actin and myosin) was measured in tissue taken from areas where muscle cells usually develop in normal embryos. The specific muscle proteins were totally absent in certain tissues (such as the cells enveloping the ribs that were supposed to develop into muscle; the results are not shown here).

a. Normal mouse   **b**. Mouse with no *myogenin*

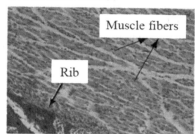 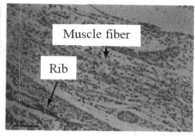

Fig. 4 Sections in the rib area of a normal embryo (**a**) as compared to an embryo without *myogenin* (**b**). In the embryo without *myogenin*, the rib muscle density is very low and only a few muscle fibers appear. On the other hand, in the normal embryo, most tissue consists of muscle fiber. The rib bone is clearly more tenuous in the embryo without *myogenin*

Discussion

This research proves that the *myogenin* gene is essential to normal differentiation of skeletal muscles in the whole organism, because mouse embryos having two copies of the mutant *myogenin* gene were born without skeletal muscle. The fact that mice with one copy of the mutant *myogenin* gene had half the quantity of *myogenin* protein in normal mice, and no protein at all was found in mice having two copies of the mutant *myogenin* gene, proves that skeletal muscles can differentiate normally with half the amount of *myogenin* but not when *myogenin* is absent in the developing embryo.

Why is *myogenin* so different from the other myogenic genes in the developing embryo? Based on the findings in this study, absence of normal *myogenin* causes severe damage to skeletal muscle. This is unlike the two other myogenic genes *MyoD* and *Myf5*: mice without *MyoD* or *Myf5* developed normally. The conclusion was that neither of these proteins alone is essential for normal skeletal muscle differentiation. Only mice with no *MyoD* or *Myf5* showed a lack of skeletal muscle differentiation. Therefore, in the differentiation process of the skeletal muscles, *MyoD* can compensate for the absence of *Myf5*, and vice versa.

Myogenin first appears in the mouse embryo several hours after *MyoD* and *Myf5*. In embryos without *myogenin*, muscles with a singular shape differentiated: in places where skeletal muscles were supposed to differentiate, mononuclear cells appeared. These cells do not fuse to form multinucleate cells and do not differentiate into muscle fiber (Fig. 1). These findings suggest that *myogenin* is not essential at the stage at which cells start differentiating into muscle cells (Fig. 1). However, in the final stages, *myogenin* is most essential, and without *myogenin* expression, the differentiation of skeletal muscle cannot be completed.

Mice without *myogenin* have severe problems in the ribs. Because *myogenin* is normally expressed only in muscle cells and not in bones, the defects found in the skeleton are probably secondary to the muscle cell differentiation deficiency: for example, there was a decrease in the forces acting on the bone by muscles; these forces may be essential to stimulating bone differentiation.

In this study, mice were born without skeletal muscle because of the absence of a normal *myogenin* gene. Future research will attempt to identify the genes that control *myogenin* expression and the genes controlled by *myogenin*.

Question

The following table presents the findings collected from several research studies in which myogenic gene expression was tampered with in the same way as described in this article for *myogenin*. Analyze the results and draw a schematic illustration to express the stage of expression of each of these genes during the differentiation process as described in Fig. 1.

| | | Phenotype | | | |
	Genotype	Primary cell	Myoblast	Myotube	Myofiber
MyoD	$(-/-)$	+	+	+	+
Myof5	$(-/-)$	+	+	+	+
Myogenin	$(-/-)$	+	+	−	−
MyoD:Myf5	$(-/-);(-,-)$	+	−	−	−
MRF4	$(-,-)$	+	+	+	−

Legend: + normal; − absent

Glossary

Cell Culture a method which allows growing cells under controlled conditions outside the living organism. Cell cultures allow performing experiments that are not possible in the whole organism, in a controlled way.

Chimera A monster from Greek mythology. Its body was made up of several types of animals; it is mostly described as having a lion's head, the body of a goat and the tail of a dragon. In research, the word chimera is used as a metaphor to describe the combination of parts whose origins are different: chimeric molecular, embryo...

Embryonic stem cells multipotent embryonic cells that can grow and multiply in tissue culture and maintain their high developmental potential.

Gain of function induction or enhancement of the expression of a certain gene.

Loss of function prevention of the expression of a certain gene or excision of the gene itself.

Transcription factor a protein having the ability to bind control regions in the DNA and specifically influence the transcription of certain genes.

Appendix B: An Example of an Adapted Primary Literature (APL) Article

Expression of the Bacillus thuringiensis (Bt) Toxin in Chloroplasts of Tobacco Plants Imparts Resistance to Insects[2]

Hedda Falk[*], Yael Piontkevitz, Gilat Brill, Ayelet Baram-Tsabari[*] and Anat Yarden

Department of Science Teaching, Weizmann Institute of Science, Rehovot, Israel

Abstract

Previous studies, in which a toxin-encoding gene isolated from the bacterium *Bacillus thuringiensis* (*Bt*) was introduced into the nucleus of transgenic plants, led to an attempt to introduce the operon encoding the toxin into tobacco plant chloroplasts. We integrated the operon into the chloroplast genome of the transgenic plants and it was expressed from this genome. The toxin protein that formed in the transgenic plant cells crystallized, in the same way that the toxin forms crystals in the *Bt* bacterium. The quantity of toxin in these transgenic plants was a 100 times higher than that in transgenic plants into whose nucleus the toxin gene had been introduced, and it is one of the highest levels reported to date in transgenic plants. The quantity of toxin remained constant even after the leaves turned yellow. When various moth larvae were fed the transgenic plants, they stopped feeding and died after a short time. These results pave the way to a new biotechnological method of creating transgenic plants.

Introduction

Insect larvae that feed on crops are one of the main problems of agriculture throughout the world. The modern solution to this problem is to introduce the toxin-encoding gene of the bacterium *Bacillus thuringiensis* (*Bt*) into plants. This

---

[2] This paper is an adaptation based upon De Cosa, B., Moar, W., Lee, S.-B., Millar, M., & Daniell H. (2001). Overexpression of the *Bt cry2Aa2* operon in chloroplasts leads to formation of insecticidal crystals. *Nature Biotechnology, 19*, 71–74.

[*] This adaptation was completed when Hedda Falk and Ayelet Baram-Tsabari were Ph.D. students at the Weizmann Institute of Science.

Fig. 1 Spore of the *Bt* bacterium (*Bacillus thuringiensis*) and the toxin crystal attached to it

Bacillus wall
Toxin crystal

Spore
Spore wall

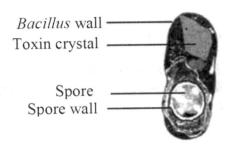

bacterium produces a protein—a toxin—that is toxic to larvae and insects. In the bacterium, the toxin turns into a crystal that binds to a **spore** created by the bacterium (Fig. 1).

The toxin in its crystal form is not toxic to insects. However, in the digestive tract of certain insect larvae (mostly from the order Lepidoptera), the pH level is alkaline (pH = 9) and, in this environment, the crystal dissolves. The soluble toxin can bind to **receptors** on cells that make up the intestinal wall of the larva. Binding between the toxin and the receptors leads to destruction of the intestinal wall. Within minutes to hours, the larva stops eating and, in hours to days, it dies as a consequence of the destruction of its digestive system and blood poisoning. The *Bt* toxin does not harm insects that have different pH levels in their digestive tract (for example, carnivorous insects), or organisms that lack the specific receptors for *Bt* toxin (for example, amphibians, fish and mammals, including humans).

In the *Bt* bacterium, the crystalline toxin is encoded by an operon. The operon includes the toxin-encoding gene and two additional genes encoding proteins that allow crystallization of the toxin. In previous studies, isolation of the toxin protein-encoding gene from the *Bt* bacterium and its engineering into the plant nucleus were described.

Plant cells into whose nucleus the *Bt* toxin gene, isolated from the *Bt* bacterium, was inserted expressed the gene. The toxin protein produced in these cells was soluble and found in all plant cells, including those in the pollen grains and roots. Experiments in the laboratory and in the field showed that these engineered plants are resistant to different insects, including *Heliothis* moth larvae (particularly harmful to cotton crops throughout the world) and European corn borer larvae—an insect against which no efficient chemical pesticide has been found.

However, engineered plants containing the *Bt* toxin gene in their nucleus present a number of problems, the main ones being that:

1. the toxin is produced in all plant cells and can penetrate the soil and harm insects living close to the roots;
2. the level of toxin expression is low, and the protein tends to degrade quickly because it is unstable in its soluble form;
3. the toxin is produced as a soluble protein and therefore can harm a wider variety of insects;
4. the Bt toxin gene can pass to other plants by means of the pollen grains.

One way to overcome these problems might be to insert the whole toxin operon into the plant, thereby allowing the formation of stable toxin crystals in the plant cells. However, because the eukaryotic genome of the plant nucleus does not allow expression of a prokaryotic operon, it is not possible to engineer the toxin operon into the nuclear genome. Studies that have attempted to insert several foreign genes into the nucleus separately, each one controlled by a distinct regulatory region, have demonstrated the complexity and awkwardness of this procedure. Therefore, no experiments have been attempted to insert all of the genes of the operon encoding the crystalline toxin into the nucleus. To overcome these problems, an innovative method was used that enables inserting foreign genes into the **chloroplast** genome. The main advantage of inserting a gene into the chloroplast is that the quantity of protein that will be encoded by that gene will be much larger than the quantity of protein encoded by a gene inserted into the nucleus. This is because in every plant cell, there are hundreds of chloroplasts, and in each one of them there are large numbers of chloroplast DNA molecules. An additional advantage is the similarity between the chloroplast genome and the prokaryotic genome, which allows the insertion and expression of a whole **operon** containing a number of genes involved in a certain biochemical process controlled by a single regulatory region—unlike the insertion of single genes. Therefore, inserting the *Bt* toxin operon into the chloroplast can enable the formation of toxin crystals in the plant cells as well.

In this study, we describe the engineering of transgenic tobacco plants that contain the *Bt* toxin operon in their chloroplasts. In these plants, both the quantity of Bt toxin and the plant's toxicity to the larvae of different insects from **the order Lepidoptera** were measured, and compared to plants in which the toxin gene was engineered into the nucleus.

Materials and Methods

Organisms

In this study, three tobacco plant varieties were used:

(a) wild-type, non-transgenic tobacco plants;
(b) engineered tobacco plants in which the toxin gene was inserted into the cell nucleus;
(c) engineered tobacco plants in which the operon for the *Bt* bacterium toxin was inserted into the genome of the chloroplast.

After engineering, the plants were grown in tissue culture under sterile conditions.

For toxicity tests, the engineered plants were fed to larvae of three moth varieties that are harmful to agricultural crops: *Helicoverpa zea* (corn earworm), *Spodoptera exigua* (beet armyworm), *Heliothis virescens* (tobacco budworm), and to the larvae of the monarch butterfly (*Danaus plexippus*), which is not harmful to agricultural crops. The larvae were grown in the laboratory under controlled conditions (Fig. 2).

Fig. 2 Larvae and adults of tobacco moth, beet armyworm moth, cotton moth, and monarch butterfly

Preparing the Vector for Engineering the Toxin Gene

Creating engineered DNA (Fig. 3): After isolating the structural region of the operon encoding the toxin from *Bt* bacteria (*Bacillus thuringiensis*), the transcription regulatory region isolated from one of the specific chloroplast genes is attached

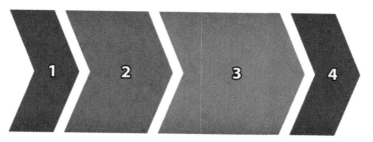

Fig. 3 Engineered DNA containing the operon for the *Bt* bacterium toxin

to it, as well as a gene imparting antibiotic resistance (originating from the bacterium *Escherichia coli*) to serve as a selector gene. The resistance gene is controlled by the same regulatory region. At one end of the toxin operon, a DNA segment that stimulates transcription is attached, also originating from the chloroplast. The two regulatory regions originating from the chloroplast, for transcription stimulation and repression, restrict the expression of the toxin operon and the resistance gene to the chloroplast. They cannot be expressed in the nucleus. The plasmid containing sequences from the chloroplast genome is used as a **vector** for the *Bt* operon. The vector can penetrate the plant cell and the chloroplasts and integrate into the chloroplast genome. The selective gene will also only be expressed when the vector is in the chloroplast.

Inserting the Operon for the *Bt* Toxin into Tobacco Plant Cells by Biolistic Bombardment

Biolistic bombardment serves to insert DNA into plant cells by means of a tool called "gene gun". Metal balls of 1 μm diameter are prepared with a coating of the vector molecules. The coated metal balls are introduced into the "gene gun" barrel and shot at the cells from short range (9 cm) at high velocity. As a result of the bombardment, some of the balls enter the cells, and some enter the cell organelles, among them the chloroplasts (Fig. 4). In the liquid environment of the cell, the vector molecules break off from the balls. Some of them integrate into the chloroplast genome and some of them into the nuclear genome. Because the regulatory regions for starting and ending transcription of the engineered operon are specific to the chloroplast, the operon will be expressed only in the chloroplast genome and will not be expressed in the nucleus, even if it penetrates it.

Isolating and Growing Plant Cells Containing the Operon for the *Bt* Toxin

To select the cells that have the foreign DNA inserted in their chloroplasts, the cells are grown under sterile conditions in the presence of antibiotics. Antibiotics are

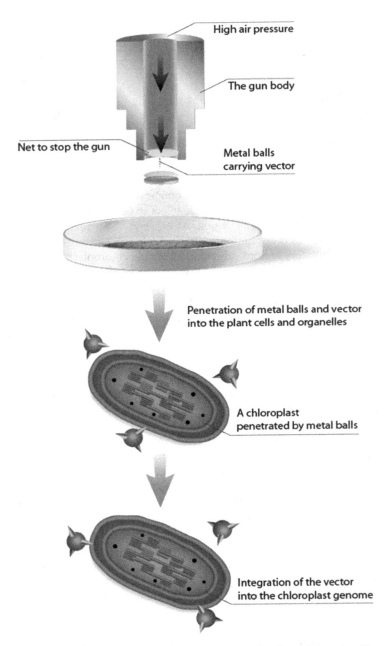

Fig. 4 Schematic description of insertion of the operon encoding the toxin into the chloroplast of the tobacco plant by biolistic bombardment

used to prevent protein synthesis in the chloroplast, and thus to prevent the development of plants that are not engineered. In other words, only engineered plants that express the gene encoding antibiotic resistance will be able to grow and develop in the presence of the antibiotics. Because the gene that imparts antibiotic resistance was inserted into the vector such that the regulatory region would allow its specific expression only in the chloroplast, only cells that contain the vector in their chloroplasts will develop into a whole plant in the presence of antibiotics.

Determining Toxin Protein Levels by the ELISA Method

The protein concentration of the *Bt* toxin in extracts of the engineered plant leaves, in which the gene encoding the *Bt* toxin is expressed, is determined by enzyme-linked immunosorbent assay (ELISA). This method, which enables the detection and quantification of various proteins (such as hormones, enzymes, toxins, etc.), includes four main stages: coating the wells with antibodies, adding a sample containing the target protein, exposing the sample to secondary antibodies attached to an enzyme, and detecting a color reaction. Between each stage, the components that do not attach to the wells are rinsed, as explained further on (see Fig. 5). The protein concentration of the *Bt* toxin is determined by means of a **calibration curve**: known quantities of *Bt* protein are added to wells which have been coated with antibodies to the *Bt* protein. A second antibody linked to an enzyme is added to these wells. After the addition of specific substrate for the enzyme, the color intensity in the wells is measured. It is possible to prepare a curve that expresses the protein-dependent color intensity. By means of this curve, it is possible to quantify the *Bt* protein in different samples where the protein quantities are not known. To perform the measurement, the non-soluble proteins in the cell, among them the toxin crystals, first have to be dissolved. Consequently, the toxin quantity is expressed as a percentage of the total soluble proteins in the cell.

Results

Expression of *Bt* Toxin in the Tobacco Plant Leaves

The transgenic tobacco plants were tested for *Bt* toxin production and quantity by means of the ELISA method (Fig. 6). In tobacco plants engineered for *Bt* toxin in the nucleus, the quantity of toxin in mature green leaves was 0.4 % of the total soluble protein in the cell. In young green or yellowing leaves, the quantity of toxin was low (less than 0.1 %). On the other hand, in plants in which the *Bt* toxin operon was inserted into the chloroplasts, the quantity of toxin in mature green leaves was high (48 % of the total soluble protein in the cell). In these plants, the amount of toxin increased with leaf age, and remained high, even when the leaves turned yellow (Fig. 6). Despite the large quantity of engineered protein in the chloroplasts, photosynthetic capability was not impaired, and the tobacco plants flowered and produced seeds normally (the results are not shown here).

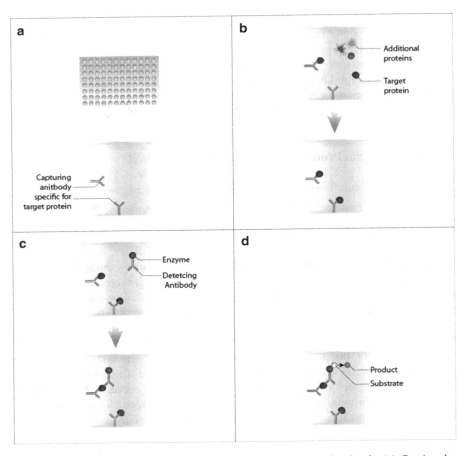

Fig. 5 Stages of the ELISA method to determine protein concentration levels. (**a**) *Coating the well surfaces with antibodies*: A solution containing antibodies that specifically bind to the protein that is to be quantified is introduced into the wells. These are capturing antibodies. The antibodies chemically adhere to the well surface such that each well contains numerous antibodies coating the surface. At this stage, the antibodies that do not bind to the surface are washed away. (**b**) *Adding a sample containing the target protein*: A sample solution containing a protein mixture is added. Only the target protein against which the capturing antibodies were prepared will bind with them. At the end of this stage, the proteins that do not bind to the capturing antibodies are washed away. (**c**) *Exposure to enzyme-linked detecting antibodies*: Other antibodies against the target protein are added to the well. Enzyme molecules are attached to these antibodies that catalyze a color reaction in the presence of a certain substrate, hence the term "detecting antibodies". These antibodies bind to the target protein at a different site. A combination is created (similar to a "sandwich"): capturing antibody – target protein – detecting antibody. At the end of this stage, the excess detecting antibodies that do not bind to the protein are washed away. (**d**) *Color reaction*: A substrate that reacts with the enzymes attached to the detecting antibodies is added. As a result, the solution color changes. Because an equal quantity of substrate is added to each well, the more enzyme molecules, the more intense the color. The quantity of enzymes in the wells is in direct proportion to the quantity of target protein attached to the detecting antibody. Therefore, the color intensity is in direct proportion to the target protein. A calibration curve is prepared by measuring the color intensity produced by the presence of a known quantity of target protein. With the help of such a calibration curve, it is possible to evaluate the quantity of target protein in the sample

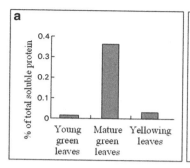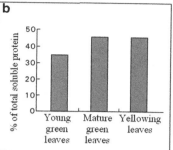

Fig. 6 Concentration of *Bt* toxin protein in leaves of engineered tobacco plants. (**a**) Plants engineered with *Bt* toxin gene in the nucleus. (**b**) Plants engineered with *Bt* toxin operon in the chloroplast

Insect Reaction to the Engineered Plants

To examine the insects' reaction to the engineered plants, larvae of three moth species that are harmful to agricultural crops: *Helicoverpa zea*, *Spodoptera exigua*, and *Heliothis virescens*, were fed leaves of the two types of engineered tobacco plants: those containing the Bt toxin gene in the nucleus and those containing the Bt toxin operon in the chloroplast. As a control, larvae were fed leaves of wild-type tobacco plants (Table 1). The three different types of larvae ate all of the leaves of the non-engineered tobacco plant. From the quantity of excrement in the dishes, it can be assumed that they digested these leaves. When the same types of larvae were fed leaves from the engineered tobacco plants containing the *Bt* toxin in the chloroplast or nucleus, they ate for short periods of time, stopped feeding and died within a few hours to a few days. Very little difference was observed between the influence of engineered tobacco plants containing the toxin gene in the nucleus and those containing the operon for the *Bt* toxin in the chloroplast. All moth larvae died within 3–4 days after feeding on engineered plants containing the operon for the *Bt* toxin in the chloroplasts, whereas larvae that fed on the plant leaves containing the toxin gene in the nucleus died within 3–8 days (depending on the moth type). *Danaus* larvae, which are not harmful to agricultural crops, fed leaves sprayed with pollen grains of the chloroplast-engineered plants, were not harmed at all (the results are not shown here).

Discussion

This study describes a new method for the transformation of a bacterium-originating operon into plant cell chloroplasts and its expression in them. Research results show that the *Bt* toxin gene penetrated the chloroplast. Because every plant cell has hundreds of chloroplasts, and each chloroplast has a number of DNA chloroplast molecules, the toxin level produced in the engineered cells was 100 times higher than that in engineered plants containing the gene for the toxin

Table 1 Influence of feeding on different tobacco plants on moth larvae

Plant type / Species	Wild-type tobacco plant	Tobacco plant containing the toxin gene in the nucleus	Tobacco plant containing the Bt toxin operon in the chloroplasts
Beet armyworm moth larva	The leaves were eaten up in 24 hours. Numerous excrements can be observed. All larvae remained alive.	A small quantity of the leaves was eaten up. The larvae stopped feeding after 24 hours and died within a further 48 hours.	The leaves remained practically untouched. The larvae stopped feeding after 24 hours and died within a further 48 hours.
Tobacco moth larva	The leaves were eaten up within 24 hours. Numerous excrements can be observed. All larvae remained alive.	The leaves remained practically untouched. The larvae died within 5 days.	The leaves remained practically untouched. The larvae died within 3 days.
Cotton moth larva	The leaves were eaten up within 24 hours. Numerous excrements can be observed. All larvae remained alive.	The leaves were partially eaten. Larvae continued feeding on the plant leaves for 3 days and died 5 days later.	A certain amount of leaves was eaten. The larvae fed for the first 24 hours, stopped feeding and died after 3 days.

in their nucleus. In addition to the increased expression caused by insertion into the chloroplast, the protein level of the toxin in plants containing the operon for the *Bt* toxin in their chloroplasts remained high and constant throughout the different developmental stages of the leaf, because the operon allows the formation of toxin crystals in the plant. These protein crystals are stable. Therefore, as compared to plants that do not contain the operon for the toxin, the protein is less sensitive to breakdown by enzymes that digest proteins inside the cell. In contrast, in plants that contained only the gene, and not the operon for toxin, the toxin was more sensitive to being broken down, and was only expressed in mature green leaves, with relatively low levels of toxicity (see Fig. 6).

When three species of moth larvae that are harmful to plants were fed the engineered tobacco leaves under laboratory conditions, they all died (Table 1). Only a small difference in influence was observed between the plants containing the gene for the toxin in their nucleus and those containing the operon for the toxin in their chloroplasts. It can be assumed that because the plants containing the operon produce a higher quantity of toxin, its influence in insect control under field conditions will be more pronounced than that of plants containing the gene for the toxin in the cell nucleus.

Insertion of the operon for the *Bt* toxin into chloroplasts solves a number of ecological problems that arose when the gene for the toxin was inserted into the nucleus of plant cells. The plants containing the *Bt* toxin operon in their chloroplasts produce the toxin as a crystal, which is toxic to a narrower range of insects. This is because the crystal has to be solubilized before the toxin can attach to the intestinal cells and cause death. The toxin crystal can only be solubilized in an alkaline environment (pH 9). This environment is typical to larvae of the order Lepidoptera that feed on agricultural crops, such as those tested here.

One of the objections to the use of engineered plants that are resistant to insects is that these plants will also harm other insects that do not damage crops. This claim is based on the results of experiments performed in the past, in which the pollen grains of the engineered plants containing the toxin gene in the nucleus harmed the larvae of the monarch butterfly *Danaus*. In the study described in this article, the *Danaus* larvae were fed pollen grains of a plant containing the operon for the *Bt* toxin in its chloroplasts, and the larvae continued feeding and producing excrement throughout the experiment. It can thus be assumed that the pollen grains of these plants do not contain the toxin. Another apprehension with engineered plants is the appearance of a foreign protein in different parts of the plant, such as fruit, seed or root cells, which may secrete it into the soil. In engineered plants containing the operon for the *Bt* toxin in their chloroplasts, inclusion of the gene's regulatory region ensures that the operon will be expressed only in the chloroplasts. An additional advantage of inserting the operon for the *Bt* toxin into the chloroplasts is that the gene will not be present in the pollen grains, and thus the risk of passing the foreign traits on to non-engineered plants is decreased. Following the successful application of this method in tobacco plants, it should be applicable to other agricultural crops as well. In addition to being an efficient solution to controlling harmful insects, inserting genes or operons into the plant's chloroplast genome can serve as a model for the production of different proteins serving a variety of purposes.

Glossary

Biolistics A method by which DNA sequences are inserted into cells, mostly plant cells, by bombarding with silver metal balls coated with DNA sequences.

Calibration Curve Body of data collected from measuring a known variable by means of a different variable. For example: measuring the absorbance by optical density of a series of test tubes containing increasing known concentrations of a certain enzyme. These data are later compared to a sample in which data on the target variable (in this example: the enzyme concentration) are unknown.

(continued)

Chloroplast A chloroplast is an organelle that contains unique genetic material which is capable of self-replication. In green cells, the chloroplast contains chlorophyll and is the site of photosynthesis. The chloroplast's DNA—the chloroplast genome—is of prokaryotic origin.

Operon A cluster of adjacent genes on the chromosome of a prokaryotic organism. Control and expression of the operon genes is collective, and for the most part, the products of the genes in an operon take part in the same biochemical processes in the cell.

Receptor A molecule, usually on the cell membrane, that can specifically bind to another molecule. As a result of this binding, a chain of biochemical reactions occurs inside or outside the cell.

Spore A minute uni- or multicellular reproductive body that breaks off of the parent and causes the formation of a new entity. Spores can be created sexually or non-sexually in plants, fungi, bacteria and protozoa, usually as a reaction to extreme changes in environmental conditions (for example, lack of food).

The Order Lepidoptera An insect order that includes butterflies and moths.

Vector A molecule serving to transfer DNA segments into cells, such as a plasmid or virus.

Index

Printed in the United States
By Bookmasters